Lecture Notes in Computer Science 6511

Pedro J. Marron
Thiemo Voigt
Peter Corke
Luca Mottola (Eds.)

Real-World Wireless Sensor Networks

4th International Workshop, REALWSN 2010
Colombo, Sri Lanka, December 16-17, 2010
Proceedings

Volume Editors

Pedro J. Marron
University of Duisburg-Essen
47057 Duisburg, Germany
E-mail: pjmarron@uni-due.de

Thiemo Voigt
Swedish Institute of Computer Science
16440 Kista, Stockholm, Sweden
E-mail: thiemo@sics.se

Peter Corke
Queensland University of Technology
4000 Brisbane, QLD, Australia
E-mail: peter.corke@qut.edu.au

Luca Mottola
Swedish Institute of Computer Science
16440 Kista, Stockholm, Sweden
E-mail: luca@sics.se

Library of Congress Control Number: 2010939946

CR Subject Classification (1998): C.2, I.2, D.2, C.2.4, I.6, I.2.11

LNCS Sublibrary: SL 5 – Computer Communication Networks
and Telecommunications

ISSN 0302-9743
ISBN-10 3-642-17519-8 Springer Berlin Heidelberg New York
ISBN-13 978-3-642-17519-0 Springer Berlin Heidelberg New York

springer.com

Printed in Germany

Typesetting: Camera-ready by author, data conversion by Scientific Publishing Services, Chennai, India
Printed on acid-free paper 06/3180

Preface

Welcome to the proceedings of REALWSN 2010, the 4th Workshop on Real-World Wireless Sensor Networks!

After three meetings in Europe we decided to hold REALWSN in exciting Sri Lanka. We want to thank the local organizers as well as the authors, attendees and members of the technical Program Committee, Demo and Poster Chairs for making this event possible.

As the name of the workshop suggests, REALWSN is a forum for people interested in real-world issues in the fascinating research area of wireless sensor networks. Despite many years of research the deployment of real sensor networks is still a challenging task. The behavior of real deployed networks differs substantially from the behavior of the same network in a simulator. The main objective of REALWSN is to bring together researchers and practitioners to understand these differences and boost the state of the art in this exciting field.

This year the program consisted of 11 full papers and five short papers carefully selected from over 34 submissions. Since REALWSN 2010 was a stand-alone two-day event, the attendees could also look forward to a poster and demo session with more than 10 contributions. The technical program covered topics from low-level communication and software development to a variety of real-world sensor network applications, some of them tailored to Asian wildlife which we think is particularly interesting.

Thanks again to all people who contributed to the workshop: the Technical Program Committee, the demo and poster chairs Kameswari Chebrolu and Adam Dunkels, the Publication Chair Luca Mottola and our sponsors that include the University of Colombo, the Uppsala VINN Excellence Center for Wireless Sensor Networks WISENET, InterBlocks Ltd. and the Sustainable Computing Research Group at the University of Colombo. The local Organizing Committee provided tremendous help that made REALWSN possible.

December 2010

Thiemo Voigt
Pedro José Marrón
Peter Corke
Kasun De Zoysa

Organization

REALWSN was organized by the University of Colombo, School of Computing.

General Chair

Thiemo Voigt	Swedish Institute of Computer Science, Sweden

TPC Co-chairs

Pedro José Marrón	University of Duisburg-Essen, Germany
Peter Corke	Queensland University of Technology, Australia

Poster and Demo Co-chairs

Kameswari Chebrolu	IIT Bombay, India
Adam Dunkels	Swedish Institute of Computer Science, Sweden

Local Organizers

A.R. Weerasinghe	UCSC, Sri Lanka
T.N.K. De Zoysa	UCSC, Sri Lanka
C.I. Keppitiyagama	UCSC, Sri Lanka
K.M. Thilakarathna	UCSC, Sri Lanka

Publication Chair

Luca Mottola	Swedish Institute of Computer Science, Sweden

Program Committee

Muneeb Ali	Princeton University, USA
Björn Andersson	Polytechnic Institute of Porto, Portugal
Jan Beutel	ETH Zürich, Switzerland
Torsten Braun	University of Bern, Switzerland
Nirupama Bulusu	Portland State University, USA
Rachel Cardell-Oliver	University of Western Australia, Australia

Kasun De Zoysa	University of Colombo, Sri Lanka
Carlo Fischione	KTH Stockholm, Sweden
Richard Gold	Ericsson, Sweden
Per Gunningberg	Uppsala University, Sweden
Wen Hu	CSIRO, Australia
Polly Huang	National Taiwan University, Taiwan
Raja Jurdak	CSIRO, Australia
Chamath Keppitiyagame	University of Colombo, Sri Lanka
Purushottam Kulkarni	Indian Institute of Technology Bombay, India
Koen Langendoen	TU Delft, The Netherlands
Hock Beng Lim	Nanyang Technological University, Singapore
Luis Orozco	University of Castilla la Mancha, Spain
Gian Pietro Picco	University of Trento, Italy
Utz Roedig	University of Lancaster, UK
Christian Rohner	Uppsala University, Sweden
Kay Römer	ETH Zürich, Switzerland and University of Lübeck, Germany
Jochen Schiller	FU Berlin, Germany
Cormac Sreenan	UC Cork, Ireland
Arno Wacker	University of Duisburg, Germany
Tim Wark	CSIRO, Australia

Referees

Mikhail Afanasyev
Markus Anwander
Jose Araujo
Piergiuseppe Di Marco
Olaf Landsiedel
Luca Mottola
Pangun Park
Stefano Tennina
Gerald Wagenknecht

Sponsoring Institutions

InterBlocks Ltd.
Sustainable Computing Research Group at the University of Colombo
Swedish Institute of Computer Science
University of Colombo
Uppsala VINN Excellence Center for Wireless Sensor Networks WISENET

Table of Contents

Applications I

OS Support and Programming

Applications II

Communication and MAC

Poster and Demonstration Abstracts

K2: A System for Campaign Deployments of Wireless Sensor Networks

Doug Carlson, Jayant Gupchup, Rob Fatland*, and Andreas Terzis

Johns Hopkins University Department of Computer Science
Microsoft Research*
{carlson,gupchup,terzis}@cs.jhu.edu,Rob.Fatland@microsoft.com

Abstract. Environmental scientists frequently engage in "campaign-style" deployments, where they visit a location for a relatively short period of time (several weeks to months) and intensively collect measurements with a combination of manual and automatic methods. We present K2, a mote-based system which brings high-quality automated monitoring to deployments of this nature. We identify key application requirements, describe the design and evolution of K2, and present performance results from two field deployments (the largest lasting $\sim$ 5 weeks and including 50 sensing nodes). Our results indicate that K2 is a viable scientific tool, achieving data yield $> 99\%$ and producing accurately time-stamped data, even in the absence of a persistently available reliable clock source. These results point a path towards WSN deployments managed by non-CS specialists.

1 Introduction

In this paper, we present the design, deployment, and performance results of K2 (Kampaign Koala), an evolution of the Koala [10] environmental monitoring system for "campaign-style" deployments. In this application space, domain scientists (in our case soil ecologists and atmospheric scientists) perform several weeks to several months of monitoring, frequently in remote and inaccessible regions. WSNs for campaign deployments face the problems common to most sensor networks: they must be energy-efficient and cope with lossy communications. However, for WSNs to be useful for this class of deployments, they must also be able to achieve a very large fraction of their intended data yield, timestamp all measurements accurately and be resilient to periods of unattended operation without a central basestation or persistent global clock source.

K2 combines a low-power collection protocol, post-mortem timestamp reconstruction system, and delta compression to meet these goals. Over the course of two field deployments (one with 20 nodes for 2 weeks and another with 50 nodes for $\sim$ 5 weeks), >99% of the intended data volume was collected and accurately timestamped. We were able to achieve energy efficiency which projects to a median battery lifetime above 900 days and storage capacity above 100 days with off-the-shelf TelosB motes and commercially available batteries.

P.J. Marron et al. (Eds.): REALWSN 2010, LNCS 6511, pp. 1–12, 2010.

The remainder of this paper is structured as follows. In Section 2 we give an overview of related systems and approaches. Section 3 outlines the key system requirements and describes our designs. We discuss our experience with two field deployments of K2 in Section 4. In Section 5 we explore the performance of the individual subsystems in detail and conclude with Section 6.

2 Related Work

Campaign deployments must achieve high yields, even if disconnection is common. Most WSN protocols for continuous data collection, such as CTP [3] and the protocol used for the Harvard volcano-monitoring system [15] rely on a continuously present data sink and suffer from reduced data yield and network efficiency when the sink is absent. Early efforts such as the Macroscope in the Redwoods project [14] suffered from low data yields. More recent work continues to show that high yields can be difficult to achieve in deployments [5].

The Suelo system [12] is also designed for campaign deployments, but the authors focus on how humans can complement computational techniques for fault detection, while we focus on how to ensure high data yields without infrastructure or reliable maintenance visits.

Due to the higher data rates demanded of it, Luster [13] incorporates dedicated storage nodes in the network to prevent losses due to space limitations. In our application domain, sensors need only be sampled at a modest rate (from every 30 seconds to every 10 minutes), so we can simplify matters with a homogeneous network in which individual nodes are fairly storage-efficient. Delta compression for sensor samples has been suggested before [2, 8]. We have not found descriptions of an implementation exactly like ours (where a pre-defined set of training data-derived "record types" indicate how many bits per-channel are available for records). Techniques such as those described by Li et al. [7] exploit the spatial correlation between data streams, while ours only exploits the temporal correlation of data within a single stream. These techniques seem highly promising, but we have not evaluated their impact on the complexity and performance of a system such as ours, and a simpler technique suffices to achieve our goals.

FTSP [9] can provide an accurate shared time frame for all nodes in the network, but we do not know how well it will work in a heavily duty-cycled network which may lack strong connectivity. Previous experience [4] suggested that a postmortem approach would deliver the desired accuracy and would be unlikely to fail in a partitioned network with unstable motes. Postmortem time reconstruction was described previously in [15], but their methodology is not tolerant to a missing global clock source.

K2 improves upon the previous Koala [10] system for environmental monitoring by incorporating Phoenix [4] for time reconstruction, delta compression, and other improvements to the data storage system (to reduce the data transferred by radio and stored in flash). This paper also presents results from field deployments of Koala and Phoenix, rather than from simulations and testbeds.

Our work shares common ground with that of Barrenetxea et al. [1] and Langendoen et al. [6] in its descriptions of the "nuts and bolts" engineering and deployment problems which can be as decisive in a deployment's success or failure as the technology in use. This paper focuses on the lessons learned from a different application space and complements these earlier works.

3 K2 Design

3.1 K2 Requirements

There are several key requirements which define campaign deployments:

Disconnection-Tolerance. Maintainers will visit the deployments according to their fieldwork schedules and no basestation is available between visits. Locations of interest may not necessarily form a well-connected network.
Scientifically Usable Data. Sensor measurements should be taken at regular intervals (not necessarily synchronously between motes), at a rate that is meaningful to the target application. The samples must be accurately placed in a single global time scale and post-deployment calibration should be supported.
Very High Data Recovery Rates. While low latency is not critical, the vast majority of data must be recovered eventually.

3.2 K2 Architecture Overview

Multiple **sensor nodes**, each made up of a TelosB mote [11], external antenna, battery, and sensor multiplexer board form the bottom tier of the system. Nodes take ADC samples from up to four external sensors at a fixed frequency, compress them, and store them in local flash. Nodes periodically exchange local time references with each other, but otherwise keep their radios off to save energy when not participating in downloads.

When the researchers' schedules permit it, they bring the **basestation** laptop to the field site. The basestation wakes up the sensor nodes, builds a centralized view of the network topology, and downloads any new data from nodes it can reach over multi-hop source-routed paths.

If an Internet connection is available, researchers upload the data from the basestation to the **back-end server**. This machine hosts an SQL database of the data collected thus far and performs the necessary translations from compressed data in motes' local time scales to physical values in the global time scale.

3.3 Functional Subsystems

Storage Subsystem. The storage subsystem is tailored to the requirements of campaign deployments. We want to record sensor measurements with the highest possible fidelity (i.e., raw ADC measurements), but we also want to ensure that the data recovery rate is loosely coupled with the rate of site visits: we don't want to lose data because the researcher couldn't make it to the field for a day.

Table 1. Delta compression example

Values at t_0	[100, 200]
Values at t_1	[101, 193]
Δ_1	[1, -7]
Space required per-channel (in bits)	[2, 4]
All Record Types	0:[2, 2], **1:[2, 5]**, 2:[5, 2], **3:[4, 4]**
Smallest Feasible Record Type	1

To achieve these goals, we added two layers to the TinyOS LogStorage stack and built a data-centric delta compression component.

At time t_k, a mote reads its sensors and calculates the difference from the measurements taken at t_{k-1}. It then uses the smallest pre-defined "record type" which can fit the delta. See Table 1 for an example of this procedure. Lossless compression is critical to maintaining **scientifically-usable data**, and delta compression is a simple way to achieve this in motes. Defining record types which correspond to the most-commonly-observed sets of field lengths allows us to save space over individually specifying the field lengths in every measurement.

K2 buffers these deltas in RAM before writing them to flash in order to reduce the 1-byte-per-record overhead imposed by the TinyOS LogStorage implementation. These space-saving measures improve **data recovery** and **disconnection tolerance** by extending the time to fill the nodes' flash and consequently reduce the frequency of site visits required to prevent data loss. We further improve confidence in **data recovery** by including a checksumming layer in the LogStorage stack. This writes a 16-bit CRC to the log every 1 KB of data and recomputes a CRC over the last 1 KB of data after every reboot.

Collection Subsystem. The data is collected with a modified version of Koala [10]. K2 differs from Koala in its use of a weighted (rather than thresholded) link selection scheme and random breadth-first download order (which favors "fresh" over "stale" link information). This approach supports **disconnection-tolerance** by quickly adapting to the basestation's location as the researcher takes it to different locations in the network (e.g., if the network does not form a single connected component). The Phoenix beacons described below also serve as LPP beacons for the wakeup procedure described in detail in [10].

In contrast to many collection protocols which assume a persistent base station and routing tree, K2 nodes maintain a low duty-cycle when there is no basestation (one transmission every 20 seconds). We support **data recovery** by building reliable delivery on top of the unreliable data stream primitive offered by Koala: each download attempt consists of the primary download of buffered data followed by a data gap-recovery phase during which the basestation re-requests data that was not received during the first phase.

Timestamping Subsystem. We use Phoenix [4] to assign timestamps to measurements in post-processing. Nodes broadcast their local time state every 20 seconds (current clock value and number of reboots since installation). Once per hour, nodes keep their radio on and log the beacons that they receive, along

Table 2. Summary of Deployments

Site	Nodes	Start	End	Sensors	Clock Source
Brazil	50	11/13/2009	12/18/2009	Air temp., Rel. Humidity	GPS, VM clock
Ecuador	20	05/22/2010	06/07/2010	Soil Temp., Moisture, CO_2	Laptop clock

with their current time state. This procedure produces a chain of references which are used to map the nodes' clocks to a global time scale after the fact.

This mechanism addresses the requirement for **scientifically-usable data**, by providing measurements in a meaningful time frame. It also provides good **disconnection-tolerance**: we require neither full network-wide connectivity nor a continuous global clock presence, as long as there is some limited access to the global clock and a modest degree of pairwise node connectivity.

Data-processing Subsystem. The data retrieved from the motes is first collected into a *preliminary* dataset in the field and is later uploaded to a database and processed into a *science-ready* dataset.

The *preliminary* dataset uses a single calibration curve for all sensors of the same type and uses only the basestation-to-mote time references to do timestamp reconstruction. This setup requires minimal configuration on the part of the field scientists, but still gives them enough information about the data being collected to adapt the deployment (e.g., by replacing or moving hardware). This approach promotes **data recovery** by identifying problems with data collection before the end of the deployment.

At the end of the deployment, we use Phoenix [4] to assign timestamps to the data[1] and per-sensor calibration curves to convert measured values to physical values.

4 Deployments

4.1 Brazil

Setup. This deployment gathered data for atmospheric scientists to use in improving their models of weather development at the Nucleo Santa Virginia research station in the Atlantic coastal rain forest near São Paulo, Brazil. Previous models were based on measurements taken in a few vertical columns, while this dataset provides a 2-dimensional mesh of temperature and relative humidity measurements at the canopy, taken every 30 seconds. With two temperature and two humidity sensors per mote, this campaign produced 5,418,074 data points over its duration.

40 nodes were deployed along a series of cables and towers approximately 30 meters above the forest floor and 10 were deployed in a transect along the ground. The footprint of the deployment area was approximately 100 m by 100 m. The deployment location was a valley with no line-of-sight to permanent structures, 6 km by jeep to permanent power and 17 km more to a reliable Internet connection. The research staff were able to visit the site every weekday,

[1] This can be done during the deployment as well, but better results are obtained if timestamp reconstruction is performed when all possible references are available.

barring weather. Following the deployment, we uniformly converted the ADC values to temperatures, and our colleagues calibrated each sensor individually to obtain sensor-specific readings from these.

Experience and Observations. We built two motes fitted with GPS receivers, which were planned to provide an accurate global clock source during the deployment. However, due to lithium battery shipping/sourcing problems, these were not available until December 9, a full 22 days into the deployment. We planned to use the basestation laptop's clock in cases such as this. However, to ease development, the basestation scripts were running in a virtual machine, which ran with a much more irregular clock than the host OS clock. While the VM clock gave poor global time, the local clock references (collected throughout the deployment) and GPS data (from the last nine days) were sufficient to assign timestamps to nearly all the measurements. This vindicated the timestamping subsystem and taught us a valuable lesson in cross-border research: *budget as much lead time as possible between the equipment's arrival and its deployment and test your backup systems rigorously.*

After the deployment, we found a few cases in which data from one mote exhibited a time offset in its data (e.g., its daily temperature peak was consistently a few minutes earlier than collocated motes). Closer inspection revealed that missing blocks of data were the cause: we assumed that samples were 30 seconds apart, so missing records shift the assumed timestamps of later samples earlier. We were able to detect these problems through the CRCs, but without a sequence number or timestamp in the records, we were not able to recover from it until the mote rebooted and reset its clock. This example highlights an important lesson in designing data storage systems: *error-detection is not the same thing as error-recovery.* We were able to download these sections correctly over a serial connection at the end of the deployment.

4.2 Ecuador

Setup. This deployment collected data for a study on soil respiration: it measured soil CO_2, soil temperature, and soil moisture every 30 seconds. Rather than blanketing an area, the deployment was made up of several widely-spread clusters of nodes: one in a pristine forest site, the other in a section of forest that had previously been clear-cut. The study site was accessible by a hiking trail from a research station in Ecuador's Yasuni National Park, which had permanent power and an intermittent satellite Internet connection.

The sensors used in this deployment added a layer of logistical problems. Their high cost limited the number that could be deployed, and their high power-consumption necessitated frequent battery replacement.

Experience and Observations. Access to *preliminary* data was introduced in this deployment. With access to this data, researchers could see when sensors were behaving erratically or operating outside of their effective measurement range and address these problems. They were also able to distinguish "interesting" and "uninteresting" locations and reposition sensors to get better measurements of the most valuable data. *Data quality should be matched to its expected* *use: in the field, it's more important to get data quickly and in a manageable form than it is to get publication-quality data.*

This brings up the issue of rapid hardware reconfiguration. We require researchers to keep track of the associations between sensors and nodes manually. If this mapping is not accurately captured, we are unable to convert from ADC measurements to physical values in post-processing. This will be impractical for larger or more dynamic deployments. We plan to build self-identifying sensors in the future which will allow nodes to record this metadata automatically. *"Mundane" problems such as efficient metadata management must be addressed in order for WSNs to be adopted as useful scientific tools.*

In K2, the basestation requests all data which the nodes have stored that hasn't been downloaded yet, and the database requests data which has been downloaded but hasn't been processed yet. The logs of two nodes became unreadable to the compression routine in such a manner that the basestation could collect the data, but the database couldn't process it. We transfer all outstanding data from the basestation to the back-end in a single file, and this file grew in size with each network download. Eventually, it grew large enough that we couldn't reliably transfer it over the poor Internet connection. While we were able to log in remotely to fix this problem, it brought two important lessons to light. *In campaign deployments, the basestation-to-back-end link should not be presumed to be reliable,* and *"fate-sharing" at any point in the data pipeline should be avoided.* In the future, we plan to conduct separate uploads for each node, and break these into smaller units to address each of these issues.

5 Results and Observations

5.1 Storage

For the evaluation of the storage subsystem, we consider the size of the sample data and the overhead required to record it. These results are from the Brazil deployment: we did not have training data for Ecuador and not all channels were populated, so the results are not as informative.

Figure 1(a) shows the per-node space savings from compression $(1-\frac{\text{size}_{\text{compressed}}}{\text{size}_{\text{uncompressed}}})$ for Brazil. The "Optimum" and "Achieved" compression assume four bits of overhead per record to identify the format at the decompressor[2]. "Optimum" assumes that each sensor's delta is represented with the fewest number of bits required to represent it: the only waste is from unused bits in the last byte.

We used two weeks of 30-second temperature and humidity data from a nearby weather station to choose the record types. In general, for m record types and n example records, we order the examples by the number of bits required to represent their first field, and put then put them into m equally-sized buckets (bucket 1 containing the first $\frac{n}{m}$ records, and so on). We assign one record type to each of these buckets, and set the width of its first channel to the largest

[2] In practice, "Optimum" requires many more than 16 record types to represent, so this calculation overestimates "Optimum"'s performance.

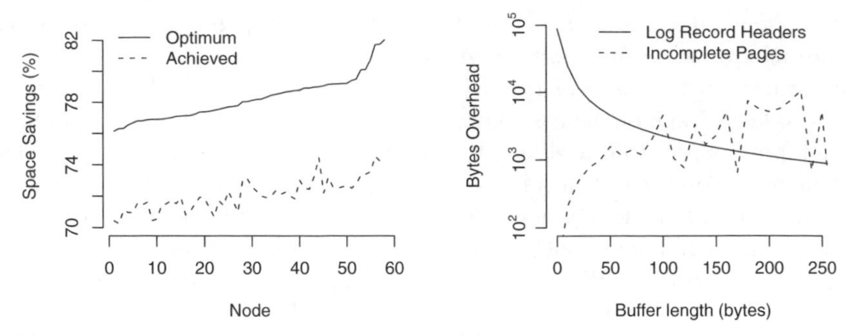

(a) Per-node space savings with compression. Node IDs reordered for clarity.

(b) LogStorage overhead. Note log scale on Y axis.

Fig. 1. Compression space savings and log overhead

first-channel width in its bucket. We then combine the buckets with the same first-channel width, and repeat this process recursively for the next channel. For example, if 50% of the records require 3 bits for channel 0, 8 of the 16 record types set aside 3 bits for channel 0. If 25% of *this subset of records* require 4 bits on channel 1, then 2 of the records would set aside 4 bits for channel 1. We had to hand-tune record types after this process to respect byte boundaries and account for the duplicated sensors.

Figure 1(a) shows that our settings achieved between 90 and 94% of optimum, even without adaptive behavior in the field or per-device customization. Adding the ability to install new settings at runtime would be a straightforward way to improve flexibility without significantly complicating the sensing node logic.

Figure 1(b) demonstrates the savings achieved by buffering data in RAM before writing it to flash. Writes can not span page boundaries, so buffer-flushes that would extend past the edge of a page waste up to buffer-length-minus-one bytes. Every write incurs one byte of "Log Record Overhead." This figure averages results from all nodes in Brazil.

In the field, we used a buffer size of 200 bytes, which worked fairly well. A shorter buffer would have reduced waste on incompletely-filled pages and decreased data lost from the RAM buffer during crashes. We recovered 99.5% of the expected data, part of the missing data is no doubt due to these lost RAM buffers. A cross-layer approach where the buffering layer is aware of page boundaries (and preemptively flushes when needed) would reduce waste on incomplete pages without increasing the overhead from log records, but this optimization did not seem worth the increase in code space and complexity to us.

5.2 Timing

Figure 2 presents the distribution of per-node timestamped data yields from the deployments. In both deployments, over 95% of the motes lost less than 1% of their data, the remainder lost less than 6% of their data. In Brazil, 99.2% of the expected data volume was assigned scientifically-useful timestamps (99.7% of the recovered data). The yield for Ecuador was comparable at 99.1% (99.4%

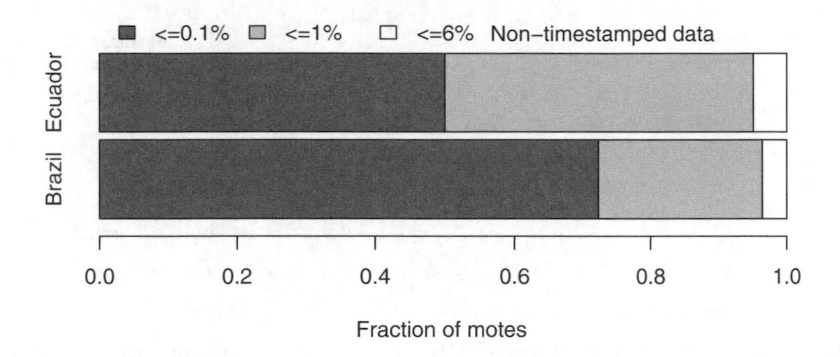

Fig. 2. Per-node data loss during timestamping

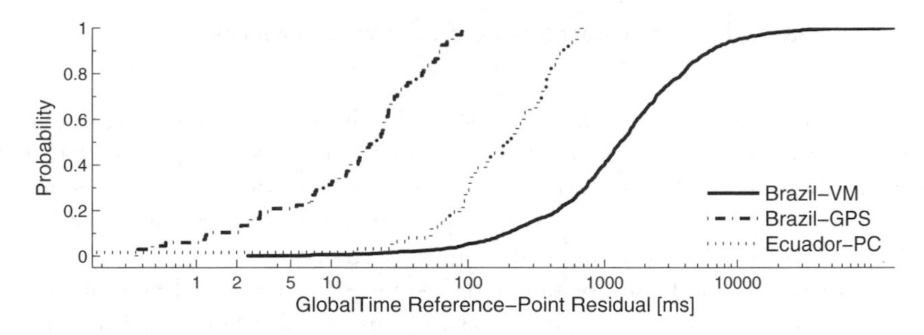

Fig. 3. Residuals of fits with global time references

of the expected data volume was collected, 99.7% of the collected data was timestamped). This is enough to satisfy the domain scientists with whom we work. In general, loss occurs when nodes reboot before they exchange enough references with neighbors to establish a mapping (due to some combination of mote instability and poor network connectivity). This was fairly rare in practice. These reboots were most likely due to transient software faults and the inability in TinyOS to place hard constraints on how often the watchdog-reset task runs.

We cannot measure the accuracy of our mechanism for assigning timestamps since we do not have ground truth. Nevertheless, we can compare the relative quality of three different clock sources. Figure 3 shows the CDF of residuals for the fits between the different global time sources and the local clock references. Low residuals indicate a good linear fit between mote clocks and the reference clock. Unsurprisingly, we see the lowest residuals for the GPS motes: these references should only deviate from a perfectly linear fit due to the effect of temperature on the mote clock and some non-deterministic delays in handling interrupts from the GPS module. The Ecuador laptop references are worse by an order of magnitude, likely due to non-deterministic delays between the time that a timestamp is put into a packet to the time that is received at the base station, forwarded over the USB, and finally timestamped in user space. The Brazil VM was another order of magnitude worse. In addition to non-deterministic

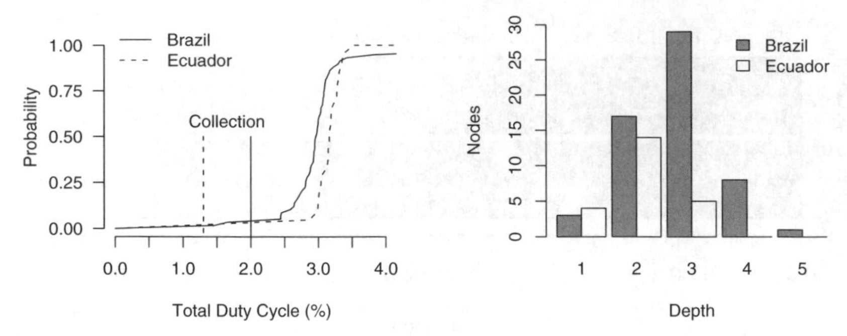

(a) CDF of per-node radio duty cycles. Vertical lines mark collection duty cycle.

(b) Node depths (median over deployment).

Fig. 4. Network characteristics of the two deployments

behavior from the host-to-VM translation, the VM clock continuously drifts and resynchronizes to the host clock, sometimes experiencing offsets on the order of seconds from the host clock's time. These experiences indicate to us that, as has been demonstrated before ([9]), removing as many layers as possible between clock references is essential to maintaining high accuracy.

Figure 4(a) shows the CDFs of per-node radio duty cycles in Ecuador and Brazil. The curves show the combined contributions of collection and Phoenix, the vertical lines show the contribution from collection alone. The median Phoenix duty cycle is 1.8% in Ecuador and 0.96% in Brazil. The main reason for this is that in Ecuador, we attempted to collect references from 10 neighbors every time that a node listens for beacons, while in Brazil we only attempted to listen for 5 before turning off the radio. In the end, this did not impact the reconstruction rate, so we will probably use lower values in the future to save energy.

5.3 Collection

The duty cycle due to data collection is 2.0% in Brazil and 1.3% in Ecuador (vertical lines in Figure 4(a)). This assumes that all nodes are active for the entire download and wakeup period. The greater efficiency in downloads in Ecuador is primarily due to the shallower and smaller network (see Figure 4(b)).

Extrapolating from these duty cycles, we can expect a node lifetime of between 930 and 1,023 days with a 19 Ah battery, which is well beyond our target lifetime of weeks or months. This could also be translated as 31-34 one-month field deployments between battery replacements[3].

The gateway doesn't necessarily reach every node on every download. Figure 5(a) shows the distribution of the number of network downloads required before a sample was received. For both deployments, more than 90% of the data

[3] This estimate assumes that 75% of the battery's total capacity is usable, and that the largest power consumer is the radio, at $\sim 20mA$ when active.

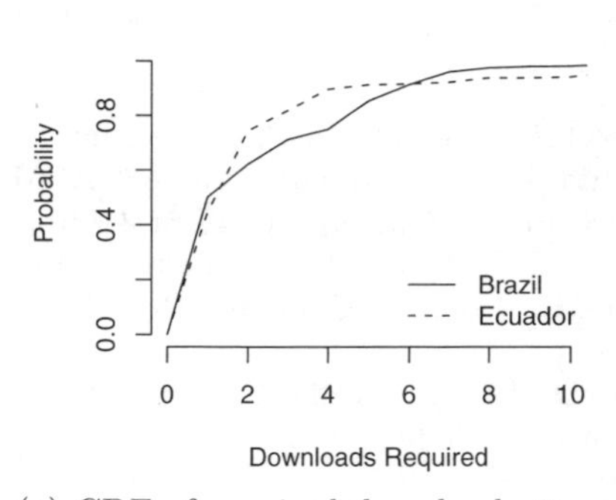

(a) CDF of required download attempts to retrieve samples.

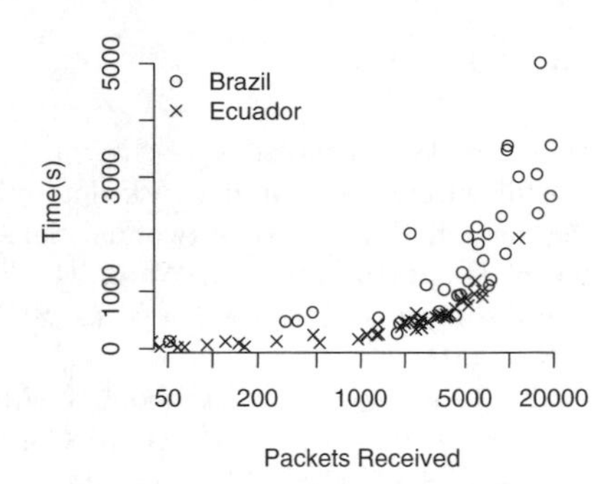

(b) Download duration as a function of data packets received. Note log scale on the X axis.

Fig. 5. Download performance

was retrieved within six downloads. In a naïve setting, this implies that site visits must occur roughly once every two weeks to safely retrieve 90% of the data. However, researchers may perform more than one download per visit, and can generally move to multiple locations to improve performance. This was the case in Ecuador, where the network layout required researchers to visit two physical locations to reach all nodes.

Our primary goal with improving storage efficiency is to prevent data loss, but this also clearly reduces the total data volume that must be transferred. By reducing data volume, we lower radio duty cycle and prolong battery life. While we can't fully separate all sources of overhead from the effect of data volume on download duration, Figure 5(b) clearly shows that the download duration is positively related to the number of data packets received, and the techniques described above will reduce the number of data packets required.

6 Conclusion

In this paper, we have described the design and deployment of the K2 environmental monitoring system and shown it to be suitable for short-term deployments of moderately-sized wireless sensor networks. Our results indicate that K2 is a viable system which can achieve high data yields, scientifically-usable results, and good battery and storage lifetimes. We have discussed how we have identified and addressed the specific challenges of remote deployments and given suggestions for future researchers in this area. In the future, we hope to improve usability and performance with automatic metadata management, more robust validation, and more effective use of the basestation's mobility.

Acknowledgments. We would like to thank Juliana Salles of Microsoft Research for project planning and Humberto Ribeiro da Rocha and his team at the University of São Paulo for field support.

References

1. Barrenetxea, G., Ingelrest, F., Schaefer, G., Vetterli, M.: The hitchhiker's guide to successful wireless sensor network deployments. In: Proceedings of the 6th ACM Conference on Embedded Networked Sensor Systems (SenSys), pp. 43–56 (2008)
2. Ganesan, D., Ratnasamy, S., Wang, H., Estrin, D.: Coping with irregular spatio-temporal sampling in sensor networks. SIGCOMM Comput. Commun. Rev. 34(1), 125–130 (2004)
3. Gnawali, O., Fonseca, R., Jamieson, K., Moss, D., Levis, P.: Collection Tree Protocol. In: Proceedings of the 7th ACM Conference on Embedded Networked Sensor Systems (SenSys), pp. 1–14 (November 2009)
4. Gupchup, J., Carlson, D., Musaloiu-E, R., Szalay, A., Terzis, A.: Phoenix: An epidemic approach to time reconstruction. In: Silva, J.S., Krishnamachari, B., Boavida, F. (eds.) EWSN. LNCS, vol. 5970, pp. 17–32. Springer, Heidelberg (2010)
5. He, Y., Mo, L., Wang, J., Dong, W., Xi, W., Chen, T., Shen, X., Liu, Y., Zhao, J., Li, X., Dai, G.: Poster: Why Are Long-Term Large-Scale Sensor Networks Difficult? Lessons Learned from GreenOrbs. In: Proceedings of ACM MobiCom (2009)
6. Langendoen, K., Baggio, A., Visser, O.: Murphy loves potatoes: experiences from a pilot sensor network deployment in precision agriculture. In: Proceedings of the Parallel and Distributed Processing Symposium (IPDPS) (April 2006)
7. Li, J., Deshpande, A., Khuller, S.: On computing compression trees for data collection in wireless sensor networks. In: Proceedings of IEEE INFOCOM (2010)
8. Mainwaring, A., Polastre, J., Szewczyk, R., Culler, D., Anderson, J.: Wireless Sensor Networks for Habitat Monitoring. In: Proceedings of ACM International Workshop on Wireless Sensor Networks and Applications (September 2002)
9. Marot, M., Kusy, B., Simon, G., Ledeczi, A.: The flooding time synchronization protocol. In: Proceedings of the 2nd ACM Conference on Embedded Networked Sensor Systems (SenSys), pp. 39–49 (November 2004)
10. Musaloiu-E R., Liang, C.J., Terzis, A.: Koala: Ultra-low power data retrieval in wireless sensor networks. In: Proceedings of the 7th International Symposium on Information Processing in Sensor Networks (IPSN), pp. 421–432 (April 2008)
11. Polastre, J., Szewczyk, R., Culler, D.: Telos: Enabling Ultra-Low Power Wireless Research. In: Proceedings of the 4th International Conference on Information Processing in Sensor Networks: Special track on Platform Tools and Design Methods for Network Embedded Sensors (IPSN/SPOTS) (April 2005)
12. Ramanathan, N., Schoellhammer, T., Kohler, E., Whitehouse, K., Harmon, T., Estrin, D.: Suelo: human-assisted sensing for exploratory soil monitoring studies. In: Proceedings of the 7th ACM Conference on Embedded Networked Sensor Systems, pp. 197–210 (2009)
13. Selavo, L., Wood, A., Cao, Q., Srinivasan, A., Liu, H., Sookoor, T., Stankovic, J.: Luster: Wireless Sensor Network for Environmental Research. In: Proceedings of the 5th ACM Conference on Embedded Networked Sensor Systems (SenSys) (November 2007)
14. Tolle, G., Polastre, J., Szewczyk, R., Turner, N., Tu, K., Buonadonna, P., Burgess, S., Gay, D., Hong, W., Dawson, T., Culler, D.: A Macroscope in the Redwoods. In: Proceedings of the 3rd ACM Conference on Embedded Networked Sensor Systems (SenSys) (November 2005)
15. Werner-Allen, G., Lorincz, K., Johnson, J., Lees, J., Welsh, M.: Fidelity and yield in a volcano monitoring sensor network. In: Proceedings of the 7th USENIX Symposium on Operating Systems Design and Implementation (OSDI) (November 2006)

TigerCENSE: Wireless Image Sensor Network to Monitor Tiger Movement

Ravi Bagree, Vishwas Raj Jain, Aman Kumar, and Prabhat Ranjan*

Dhirubhai Ambani Institute of Information and Communication Technology, Gandhinagar, India - 382007
{ravi_bagree,vishwas_jain,aman_kumar,prabhat_ranjan}@daiict.ac.in

Abstract. Wireless Sensor Network (WSN) in combination with image sensors opens plethora of opportunities in the wildlife tracking. It provides a glimpse into previously unseen, remote and inaccessible world of some of the most endangered species on earth. tigerCENSE [1]is such an attempt to put sensor network technology in conserving one of the rarest and most elusive big cat species. The node, triggered by the Passive Infrared (PIR) sensor, captures the image of tiger using a CMOS image sensor and stores it in an external memory chip. To avoid any disturbance to animal, the node uses an Infrared (IR) flash, instead of white flash, to illuminate the target at night. The stored images get transferred to the base station via radio transceiver. This is transferred to the database server through Internet links for analysis by wildlife researchers. A solar energy harvesting system for recharging node's batteries is being added to avoid frequent human visit to change the batteries, making it highly non-intrusive system.

Keywords: Camera Trap; Wireless Sensor Network; Image Sensor Network; wildlife tracking; Intrusion detection; CMOS camera; IR flash; image sensor.

1 Introduction

Most of the WSN applications have depended on sensors such as light and temperature etc., which produce small amount of data per sample. However, in recent years, technological advances, especially in CMOS, have made it possible to have very small, low powered and cheap image sensors integrated to WSN, enabling us to collect valuable visual information of the target object and its surroundings. These image sensors produce large data per sample based on image size. Due to this Wireless Image Sensor Network (WiSN) has emerged as a new field with its own application areas as well as challenges. One of its most promising applications is monitoring wildlife species.

* Corresponding author.

[1] This project is partially funded by Wildlife Institute of India, Dehradun.

P.J. Marron et al. (Eds.): REALWSN 2010, LNCS 6511, pp. 13–24, 2010.

Traditional methods of wildlife monitoring are largely based on statistical methods and data collected by ground surveys [1]. Though these methods usually yield extensive data for a given animal and its habitat, they are time consuming, expensive and unauthenticated. Some methods, such as the traditional pugmark census, are not even reliable enough[2]. Above all, most of the endangered or critically endangered species live in remote, arid and inaccessible landscape. Monitoring them, their behavior, their status and distribution becomes life threatening. tigerCENSE is an attempt in this direction to make such tracking more authentic, automated, non-intrusive, less expensive and safe. Under tigerCENSE project we are primarily focusing on collecting images of tigers along with its time/date and location to identify their movement patterns and making it available to the researchers in an easy manner.

Tiger is the largest of all the Asian big cats and one of the most threatened species [3]. Throughout their range in Asia tiger populations are threatened, either directly from poaching, or from habitat and prey loss [4]. Once having the population count above one million is now struggling for its survival with the mere population of 3,402-5,140 across the world [4]. In last few decades number of conservation programs have been proposed by various countries and other international organizations. They have been working on many possible solutions like restoring habitat, monitoring populations, anti-poaching laws etc, and millions of dollars have been invested for the same [5]. But common to all the solutions is monitoring the status and population distribution information of the tigers.

2 Available Technology

The advent of advance camera trap technology has revolutionized conservation plans for wildlife. It helped to uncover invaluable information about rare species and their habitats, which can be shared with local governments when making land-use decisions, anti-poaching activities etc. Most of the available camera traps use independent commercially available camera modules, that may be digital or film-based usually triggered by a motion detector.

In the very old days trip wires and pressure pads were used to trigger cameras [2]. Modern motion detectors are based on infrared and may be active or passive. Active infrared based motion detectors send out an infrared beam to a receptor located some distance away. When any object obstructs this beam's path, the detector triggers and camera captures the photograph. Whereas a PIR sensor tracks heat change in the surrounding. When any infrared emitting object passes in front of the detector, it detects the motion. Also, the modules that aim for night photography usually come equipped with either white or infrared flash. Some of the commercial cameras use almost 64 LEDs making them much bigger and consume lot of power.

These systems usually use strobes and wires to interconnect the motion detector, the independent camera and to setup an automatic image capturing system. This makes their size quite bulky and difficult to camouflage making it highly prone to stealing by people or being damaged by animals. Also, presently most of the commercially available traps do not have a local wireless network link. Although few of the traps do communicate with satellite but because of the leased satellite link, they cost heavily.

In this paper we are proposing the system, tigerCENSE, which has been able to resolve many of the problems faced by these traditional camera traps. Here we will be discussing the hardware and software design architecture of the tigerCENSE system at the node, base and network levels. In particular, the paper embodies the issues and constraints, which were met during the design and testing of the system.

Section 3 of the paper discusses the design parameters taken into account for tigerCENSE. Overview of the system is described in Section 4 covering the hardware and software aspects of the system. Experience gained, system performance and our field testing results are covered in Section 5. Finally we conclude by enumerating the challenges and experience gained from inception to trials.

3 tigerCENSE

To make an informed decision, researchers need to know the status and distribution pattern of tigers in the area of interest. They collect information using pugmark, DNA technique or through camera traps. As human beings have fingerprints as their unique identity, characteristic stripe patterns on cat's body differ from one individual to another and from one side of the cat's body to the other [6]. In fact, there are no tigers with identical markings. Wildlife researchers are mainly interested in these unique stripes pattern. It allows them to extract potential information on the presence of species, their home range sizes, individual recognition and density estimates, activity cycles, behavior, seasonal variation in movement and abundance and also allows for comparisons to be made between areas [2].

tigerCENSE is an attempt in the similar direction and provides images in an inexpensive, power and time efficient manner. Nodes are setup by researchers along each bifurcation of the tiger trail to help figure out the path taken by tiger. Whenever a tiger gets in the field of view of PIR sensor, an interrupt is generated and the image sensor will capture the photograph. As tiger moves mostly in night, an infrared flash is integrated in the system. The photograph is time stamped and gets stored on a micro-SD card along with node ID. Once the communication with gateway or next hop neighbor is available, it would transfer the image wirelessly using a radio transceiver. As the memory size of the micro-SD card can be increased, the upper limit of photographs that can be saved is adjustable. Also, wireless connectivity and solar recharging for battery, help in minimum anthropogenic disturbance, which otherwise would have been required for data collection and to change the power battery.

Though camera traps technology have been in use for quite long time but still it is not fully explored and suffer some major drawbacks. Besides having all the pros of old traps, tigerCENSE has been designed keeping the following drawbacks, explained further, as its prime design challenges.

Response Time. The time delay between PIR interrupt and capturing the photograph is very critical. Because of large response time of many traditional traps, fast moving animals do not get captured. tigerCENSE system needs to reduce this time to around one second to overcome the said drawback.

Size and Cost. Traditional traps were bulky and costly. Developing a customized system with an integrated image sensor is desirable as this will drastically reduce the size and the cost. Also, presently the number of LEDs used to illuminate the animal is very high. We could make an illuminator with fewer but brighter and more efficient LEDs.

Disturbance to Animals. To allow night photography IR flash is recommended as white flash will startle the animal resulting in the abandonment of the path. Also, mechanical shutter produces a click sound, while taking a photograph. This needs to be avoided as this makes the animal cautious of its surrounding and to behave abnormally.

Automated Data Transmission and Local Storage. A wireless connectivity is required which allows the nodes to be deployed in very remote areas and it will also reduce human visit to the forest to a great extent. Also, to compensate for any link failure due to environment or other failures, the node should have sufficient external memory to store the data for a month.

Remote Configuration. Researchers need to go to the field each time they need to change any parameter, like number of shots in burst mode, delay between two adjacent shots etc. of the traps. Remotely changing of parameter further reduces the visit and labor of researchers.

Fail-Proof against False Interrupts. In spring when many trees shed their leaves in preparation for new foliage, active IR sensor gives lot of false interrupts. Each momentary break in the beam caused by a leaf floating across the path may result in a useless picture being taken. tigerCENSE needs to take care of it as this may consume large amount of power for no good reason.

Health Information of Traps. Presently, once the trap is deployed in the field, there is no way to know about its health and other parameters. The film or the battery might have been exhausted long back but the researchers would not know. Also, the camera might stop working because of some technical problems, it might get stolen or may have been damaged by an animal, but it will remain unknown until someone visits it. This makes the trap highly inefficient as it may loose important information.

Energy Harvesting. Present traps consume enormous power and need the battery replacement at regular interval. This not only leads to frequent visit of the researchers but also the maintenance cost goes up. This requires for an efficient power supply with a recharging mechanism. The Solar recharging system could be an excellent solution to it. With careful energy management policy, supplemented by harvesting, the energy requirements can be met.

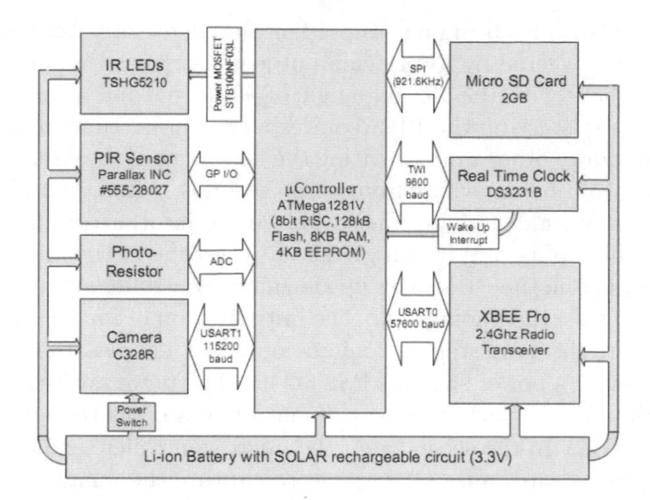

Fig. 1. tigerCENSE Hardware setup depicting various components, their interfacing and power supply

4 System Overview

Broadly the tigerCENSE system is divided like any other traditional WSN in the hardware, related system software and drivers, middle-ware servers with data logging and web hosting services and finally the browser based visualization software. We plan to use radio transceivers capable of communicating over 1.5 km in the free space. The range may get affected due to surroundings but would still be sufficient to allow node to node communication and multihopping of data. Most of the nodes would be in the valley but to provide link to Internet, we would need to use GPRS links using mobile communication infrastructure. As mobile signals would not be available in the valley, we would need to setup a 4-5 km directional link between a gateway in the valley and that on the hills. Using mobile signal available on the hills, we would be able to transfer the data to servers using GPRS. Our focus in this paper is more on the node development and not on the rest of the system, where standard existing technology can be used.

This section describes the platform developed and used for our experiments. Hardware system architecture of tigerCENSE node is as depicted in Figure 1. While describing the hardware used, we will also discuss the flow of the software and the challenges faced during its development.

When the system is in idle state with no movements of animal, all the hardware components will be in power saving or sleep mode except the PIR sensor. When an intrusion is detected PIR sends an interrupt to the micro-controller and the system gets into its active state. The PIR Sensor is a pyro-electric device that detects the motion by measuring changes in the infrared levels emitted by surrounding objects.

PIR. We use Parallax INC #555-28027[8] PIR sensor, which works from 3.3 to 5V and draws less then 100 μA current. Also, it is less prone to false triggers, when compared to active beam interrupted motion detectors. Active beam based system may get triggered by a very small object(e.g. leaves falling of a tree). It has the Fresnel lens with the viewing angle of 90 degree and a range of approximately 20 feet. At start-up the PIR requires a 'warm-up' time in order to learn its environment or in other words creating the heat map of the environment. This start-up time could be anywhere from 10-60 seconds. After this, whenever PIR sensor detects any sudden change in its heat map, in other words it detects an intrusion; it pulls up its output pin giving an interrupt to the micro-controller.

The interrupt from the PIR wakes up the micro-controller and it initializes the image sensor to take the photograph. The initialization of image sensor happens in two steps. In the first step the micro-controller enables the power to the image sensor using a power switch TPS2092 [9]. The power switch is being used to conserve the power which otherwise would be wasted as the quiescent power of the image sensor. In the second step the micro-controller sends commands to the image sensor to customize setting and to capture the image.

Image Sensor. COMedia Ltd.'s C328R [10] image sensor module is used, which performs as a JPEG compressed, low cost, low powered still camera. It interfaces with the micro-controller using the serial communication. It works on 3.3V with 60mA of current. As we are using IR flash to illuminate the object, we use a lens without IR filter. CMOS image sensors are typically sensitive to 1000 nm and use of IR LED in 850 nm to 950 nm range to illuminate the target is possible. The lens configuration can also be altered to vary the Field of View (FOV) of the camera [11]. Currently, we are using the lens with FOV of 60 degree.

Before taking the photograph the micro-controller reads the output of a photoresistor, interfaced to its ADC pin, to sense whether the ambient light is sufficient for the image or if flash is required. Depending on the need, micro-controller switches on the high intensity Infra-Red Flash using a power MOSFET.

All the photographs need to be time stamped along with the node ID. To keep track of time on the node, we are using a Real Time Clock (RTC). When the node is powered on for the first time, it needs to be in the range of a base station to synchronize with the system time. Once the time is set, the battery backed RTC keeps the timing information for years and corrects any drift each time node communicates with the base.

Real Time Clock (RTC). We use DS3231[12] as RTC, which is one of the industry's most accurate RTC. Its power consumption is 110 μA at 3.3V. It has integrated temperature compensated crystal oscillator (TCXO) and I^2C interfacing.

A radio transceiver has been used to transfer the collected photographs and other data/health information of the node to the gateway/base station for onward transmission to the server.

Radio-Transceiver. Communication module XBee Pro[13] from Digi-Key is used, which is based on ZigBee/IEEE 802.15.4 standard. It operates at 2.4 GHz (only freely available ISM band in India), providing a range of more than a kilometer. Its RF data rate is 250 Kbps. While using this frequency results in higher power consumption for same range compared to 900 MHz, we gain in terms of much higher data rate and smaller compact antenna. Low cost, low power and ease of use are among the other advantages. It also provides five sleep modes to meet various needs of different applications. We use lowest power sleep mode as it is not a time but power critical system. Recently introduced, XBee Pro 2.5 version supports multihop transfer of data.

The image can be transferred using multihop facility provided by XBee Pro 2.5. But there are chances, because of bad weather or some other technical problem, establishing a communication link is not always possible for a sensor node especially those deployed in remote areas. So the captured image needs to be stored in some storage device. Typically the size of a photograph is 60KB. So we cannot use an internal memory and need an external storage.

Micro-SD Card. We have used micro Secure Digital (SD)[14] card, commonly used in mobile phones, which can be interfaced with micro-controller using SPI bus. The card can be manually removed and the images can be transferred into a computer, phone or even a digital camera for viewing. The conventional method of writing data into external flash memory restricts the user from viewing the images with such ease. The storage capacity of the micro-SD card is adjustable depending on the activity of the animal at the location. Currently we are using a 2GB card.

All the decision making and controlling of components on the node is done centrally by the micro-controller.

Micro-controller. ATMega1281V [15], with 128K bytes program memory, is the core processing unit of our design. It has 4K bytes of EEPROM and 8K bytes of SRAM. The availability of 2 USART ports enables independent communication of Camera and Radio transceiver with the core processing unit. The internal resonator is not accurate enough for serial communication, so an external crystal of 1.83728 MHz is used. (Limiting baud error to zero percent [15]).

An efficient energy power supply and management policy has been designed to achieve true non-intrusive nature of tigerCENSE. Energy efficiency is achieved by using very low loss DC/DC converter and other components such as power switch to switch off all the devices, whose sleep mode power consumption is not sufficiently low. All the peripherals are switched off or kept in sleeping mode, except PIR sensor, in normal mode. The system is powered by a re-chargeable Li-poly battery. Solar energy harvesting is being added to further enhance the node life. The battery's capacity should be sufficient enough to power the node for at least one month. We are carrying out tests to determine node's actual life time in working environment.

Battery. We are using a 6AH Li-poly battery[16]. These are very slim, extremely light weight batteries based on the new Polymer Lithium Ion chemistry. Its output voltage is 3.7V with 2.7V cut-off voltage. Also it has 2C discharge rate.

Designing a simple power supply for such complex system was a challenge. All components and sensors were carefully selected to have low energy consumption profile and almost similar input supply range with 3.3V as the common voltage. The decision of using a common voltage (3.3V) not only made the power supply for the node simple but also saved energy, which otherwise, would have been wasted in regulating it for different voltages. With time the battery voltage will reduce from 3.7V to 2.7V. But the node needs a constant voltage supply of 3.3V so we need a buck-boost DC converter to regulate the battery voltage.

Buck Boost Converter. To utilize the battery power to the maximum, a DC/DC converter, TPS63001[17] buck boost converter from Texas Instruments, is used. It provides a constant 3.3V output with a maximum of 1.8A of current; being rated up to 96% efficient.

The same battery will be used to power the IR LEDs. These LEDs will be used in pulse mode with high time of 30ms. To get high intensity rays, we need to supply very large current (approx. 3.0 A) for this pulse duration . As a battery may not supply such large current, we need buffer storage of electric charge in between. Super-capacitor is the best option for this task. We are using two super-capacitors in series to get the required voltage.

Super-Capacitor. Super-capacitor TS12S-R[18] is used, which is highly compact and high density capacitor with capacity of 10F at 2.5V. Its self discharge rate is very low and can supply maximum of 4.5A of current.

To switch on the LEDs for such a short time, we need a Power MOSFET with very small ON time resistance. ON time resistance is of particular importance as we are drawing very high current of 2.5A. Even few milliohms of ON resistance can result in significant voltage drop across LEDs, which will reduce its intensity severely.

Power MOSFET. The Power MOSFET STB100NF03L[19] from ST Microelectronics has been used for the said task. Its ON-resistance is less than 3.2 $m\Omega$ with Gate threshold voltage is as low as 1.7V.

Tigers mostly move in night time and to illuminate the animal, we are using IR LEDs. But since the size of the trap is of much concern, we need to use least no. of LEDs possible. This requires very high radiant intensity, low forward voltage LEDs.

IR LED. We use TSHG5210[20], which is the strongest high intensity IR LED available in the market from Vishay Semiconductors. This is an infrared, 850nm emitting diode with forward voltage of 1.5V. In pulse mode its radiant power is 2300mW/sr. Its angle of half intensity is +/- 10 degree. However, one may need wider beam angle than what this provides.

Right now the system uses 12 LEDs in parallel. We are working on the ways to reduce the number of LEDs to about 5-6 by improving the charge buffer system. We selected parallel configuration as it is easy to provide a large current instead of high voltage. Also, in such configuration each LED is independent of the other and failure of one LED does not disturb the function of whole flash.

Learning from the experience for wildCENSE[21] project the node has been designed employing numerous noise reduction techniques. To reduce the ADC noise, a LC filter (L=10mH and C=$0.1\mu F$) has been added to the ADC pins of the micro-controller. Also, the AVcc is connected to the main power supply

Fig. 2. tigerCENSE node, Front and Rear view

without any in between fan out lines, to reduce noise [22]. The whole PCB has copper pouring to keep the noise at a minimum level as also to dissipate any heat generated by the node. Figure 2 depicts the PCB made for the node. The size of the populated PCB is 3.8 x 5.6 x 3.1cm^3, weighing only 43 gms excluding power supply and enclosure.

5 Experimental Results

Based on the expected speed of movement and width of walkways(assumed 10 feet) and distance of node from walkway to be 10 feet, a delay of 1.8 sec is kept

Fig. 3. Prototype box used for testing

Fig. 4. Photograph clicked using IR Flash in dark night

between the PIR interrupt and capturing a photograph of the object. Minimum delay achievable seems to be 250 ms. It is extremely small time as compared to the response time of traditional traps which ranges into few seconds. Also, minimum delay between two continuous shots has been found out to be 1s. It is dictated by the time to transfer the data from Image Sensor to Micro-SD card and can be reduced by buffering it in a fast memory, if one needs to collect a burst of images.

To find out the minimum suitable ON time for the IR flash to capture the stripes clearly on its body, we deployed a prototype box, as shown in Figure 3, near the cage of a tiger in Kankaria Zoo, Ahmedabad, Gujarat. We programmed the node to take pictures with increasing ON time starting from 10ms to 70ms with an increment of 10ms. Figure 4 shows some of the photographs taken by the node in the dark using an IR Flash. From the experiments we concluded that an ON time of 30ms is sufficient to get a reasonable quality image with clear stripes. We need flash time to be as low as possible to reduce blur due to motion. Some commercial digital cameras use 125 ms flash time, which leads to significant blare to the extent of image being useless.

6 Conclusion

This paper presents an operational prototype for wildlife monitoring using WiSN. tigerCENSE is compact, non-intrusive, energy efficient and reliable sensing device. It not only has all the capabilities of traditional traps but has also addressed most of the drawbacks of them. Integrated development has led to minimum delay of 250 ms. The software protocols and the hardware implementation have all been carefully crafted to optimize the systems energy requirement. Further, utilizing the solar recharging mechanism, node lifetime would be enhanced.

In future, we can also add some micro-climatic sensors in order to collect ambiance information. Also, to reduce the amount of wireless data transfer, we can deploy in-situ digital signal processing technique. This will help us save both power and time which is highly crucial for the success of the system.

tigerCENSE has been mainly developed to help in the research and conserving tigers. Besides the use for conducting a census, camera traps can be very useful for many management tasks. It can be used for human surveillance as well. In the past, traps have photographed poaching parties. Although due to latency in collecting the photograph target animal prey were not saved but it eventually led to the arrest and conviction of known offenders.

Acknowledgment

We would like to thank Bharat Jethwa of GEER foundation (Gandhinagar) for always being available when needed. Discussion with WII researchers P R Sinha (Director), S P Goyal, K Sankar, B Pandav, Q Qureshi and others have been very helpful. We would also like to thank R K Sahu(Superintendent) and others of Kankaria Zoo, Ahmedabad for giving us permission to carry out trials as well as helping in the process. We would also like to acknowledge tremendous contribution made by earlier team members of tigerCENSE, especially Amrit Panda, Rigveda Kadam, Dheeraj Kota and Hemant Kavadiya.

References

1. Yasuda, M., Kawakami, K.: New method of monitoring remote wildlife via the Internet. Ecological Research 17, 119–124 (2002)
2. Nath, L.: Camera Trap in Conservation, http://www.nfwf.org/AM/Template.cfm?Section=Home&TEMPLATE=/CM/ContentDisplay.cfm&CONTENTID=8749
3. http://www.panda.org/what_we_do/endangered_species/tigers/
4. http://www.iucnredlist.org/details/15955/0
5. Staving Off Extinction: A Decade of Investments to Save the World's Last Wild Tigers (1995-2004), http://www.nfwf.org/Content/ContentFolders/NationalFishandWildlifeFoundation/ConservationLibrary/ProgramEvaluations/Staving_off_Extinction.pdf
6. McDougal, C.: The Face of the Tiger. Rivington Books, London (1977)
7. http://www.panda.org/what_we_do/endangered_species/tigers/tiger_solutions/
8. PIR Parallax 555-18017 Datasheet, http://www.parallax.com/detail.asp?product_id=555-28027
9. Texas Instrument TPS2092 Datasheet, http://www.ti.com/lit/gpn/tps2092
10. COMedia Ltd's C328RS User-Manual, http://www.electronics123.net/amazon/datasheet/C328R_UM.pdf
11. Lens of camera, http://www.electronics123.net/amazon/datasheet/C328R.pdf

12. DS3231 RTC Datasheet, http://www.maxim-ic.com/quick_view2.cfm/qv_pk/4627
13. XBee-PRO OEM RF Modules Product manual, http://www.maxstream.net/products/XBee/product-manual_XBee_OEM_RFModules.pdf
14. micro-SD Card Datasheet, http://www.sparkfun.com/datasheets/Prototyping/microSD_Spec.pdf
15. Atmel ATMega1281 Datasheet, http://www.atmel.com/dyn/resources/prod_documents/doc2549.pdf
16. Polymer Lithium Ion Batteries 6Ah Datasheet, http://www.sparkfun.com/datasheets/Batteries/UnionBattery-2000mAh.pdf
17. Texas Instruments TPS63001 Datasheet, http://www.ti.com/lit/gpn/tps63001
18. Suntan Super-capacitor TS12S-R Datasheet, http://www.sparkfun.com/datasheets/Components/TS12S-R.pdf
19. ST microelectronics Power MOSFET STB100NF03L Datasheet, http://www.st.com/stonline/products/literature/ds/9307.pdf
20. Vishay Semiconductor IR LEDs TSHG5210 Datasheet, http://www.vishay.com/docs/81810/tshg5210.pdf
21. Jain, V.R., Bagree, R., Kumar, A., Ranjan, P.: wildCENSE: GPS base Animal Tracking System. In: International Conference on Intelligent Sensors, Sensor Networks and Information Processing, Sydney, December 15-16 (2008)
22. Innovative Techniques for Extremely Low Power Consumption with 8-bit Microcontrollers,
http://www.atmel.com/dyn/resources/prod_documents/doc7903.pdf

Motes in the Jungle: Lessons Learned from a Short-Term WSN Deployment in the Ecuador Cloud Forest

Matteo Ceriotti[1], Matteo Chini[2], Amy L. Murphy[1],
Gian Pietro Picco[2], Francesca Cagnacci[3], and Bryony Tolhurst[4]

[1] Fondazione Bruno Kessler—IRST, Trento, Italy
{ceriotti,murphy}@fbk.eu
[2] Dip. di Ingegneria e Scienza dell'Informazione (DISI), Univ. of Trento, Italy
matteo.chini@gmail.com, gianpietro.picco@unitn.it
[3] Edmund Mach Foundation—IASMA, S. Michele all'Adige, Italy
francesca.cagnacci@iasma.it
[4] Biology Division, School of Pharmacy and Biomolecular Sciences, Univ. of Brighton, UK
bryonytolhurst@live.co.uk

Abstract. We study the characteristics of the communication links of a wireless sensor network in a tropical cloud forest in Ecuador, in the context of a wildlife monitoring application. Thick vegetation and high humidity are in principle a challenge for the IEEE 802.15.4 radio we employed. We performed experiments with stationary-only nodes as well as in combination with mobile ones. Due to logistics, all the experiments were performed in isolation by the biologists on our team. In addition to discussing the characteristics of links in this previously unstudied environment, we also discuss the lessons we learned from operating under peculiar constraints in a peculiar deployment scenario.

1 Introduction

Wireless sensor networks (WSNs) are applied in many scenarios, each with unique characteristics in terms of connectivity. Assessing the specifics of a target environment is usually complex, and often entails a preliminary pilot deployment.

Application context and motivation. In this paper we report about such a pilot deployment, which took place in the cloud forest of the North-Western slopes of Ecuadorian Andes during March 29–April 3, 2010, and whose details are provided in Section 2.

The work described here is part of a larger research effort targeting the monitoring of biodiversity in community-based primary cloud forest reserves in this Andean region. Indeed, this area is at the confluence of two of the world's hottest biological hotspots: the Chocó-Darién Western Ecuadorian and the Tropical Andes. Available checklists of vertebrates likely miss most reptile and mammal species, including medium-to-large ones. The knowledge about these species' use of space and community interactions is essential to ascertain their susceptibility to environmental changes and guide conservation measures. Available information is extremely sparse and based on discontinuous observations and occasional surveys. Direct observation of animals is not a robust method, due to the very dense vegetation, while traditional indirect methods, such as capture-mark-recapture or radio-tracking are extremely effort-demanding as these areas are secluded. Recent advancements in wildlife studies, e.g., the use of GPS devices, are expensive and therefore applicable to a small number of species and sample size. WSNs

P.J. Marron et al. (Eds.): REALWSN 2010, LNCS 6511, pp. 25–36, 2010.

provide a new, exciting option in such challenging environmental conditions, especially for long-term monitoring. Advantages include the need for only a single capture (to fit the node) and the possibility to study a large sample thanks to the relatively low equipment and deployment cost. However, an essential step in seizing this opportunity is the evaluation of the node performance in the target environment.

The envisioned WSN application will encompass nodes permanently deployed in the environment at known locations as well as attached with collars to the animals themselves. We intend to use motes functionally equivalent to Moteiv's TMote Sky [3], arguably the most popular platform today. However, the 2.4 GHz band used by the CC2420 radio chip on these motes is known to be highly sensitive to foliage and water—essential ingredients of a cloud forest. Therefore, the primary motivation behind the study described here was to assess the connectivity characteristics of the target environment to determine the feasibility of our WSN architecture and guide its design.

Related work. A few real-world deployments focus on forests [5], but with characteristics different from ours. Despite the importance of understanding the connectivity of the environment targeted by a WSN, this information is rarely reported in the literature. Instead, the problem is usually tackled with studies targeting either static [4] or mobile [1] scenarios. All the reported works, however, leverage the possibility to progressively refine the investigation based on the findings. Our need to define a priori the entire experimentation pushed us towards a more general methodology, something still not available in the literature. To design our study we leveraged our prior expertise in comparing the network characteristics of a tunnel against the vineyard environment [2]. However, the differences in the application scenario, involving mobile nodes, and the inability to access the experiment site demanded a significant revision of our techniques.

Challenges. The deployment itself presented non-trivial logistical difficulties due to the geographical distance and the harshness of our target environment. Things were further complicated by the fact that the WSN experiments were "piggybacked" on the biologist's trip to Ecuador for other research purposes.

As a consequence, we faced rather unusual requirements. In the literature, similar experiments are typically run by the WSN developers, often in rather controlled environments. Instead, in our case the experiments had to be run by the biologists, and *in isolation*. Remote WSN configuration was not an option, due to the absence of data connectivity from the experiment location—the jungle. Similarly, a multi-phase deployment, where the output of one experiment guides the setup of the next, was also not an option due to the distance between the experiment location and the closest Internet access, and to the duration of the experiments. The latter was limited by the biologist's already-established trip schedule, further reduced by the inevitable lost baggage.

Simply put, this meant that our hw/sw WSN setup had to work out of the box for the entire duration of the experimental campaign, and had to be simple enough to be operated by someone without expertise with this technology.

Contributions and findings. The details about our cloud forest experiments are provided in Section 3. The main contributions of this paper are the following:

1. *Low-power wireless in the jungle environment.* In Section 4 we analyze the gathered data. The depth of the analysis is somewhat limited by the aforementioned logistic problems, as we did not have a second chance to investigate the source of unexpected behaviors. However, we are not aware of other studies investigating

low-power wireless communication in an environment similar to ours and therefore, even with these limitations, we believe our study can be of value for the research community. Moreover, some of our findings are somewhat surprising. For instance, we expected links to be rather short and unreliable, due to foliage, water, and humidity. Instead, our data show that 30-meter links are common, and in some cases reliable communication occurs up to 40 m.

2. *Mobile nodes as a connectivity exploration tool.* The inclusion of experiments with mobile nodes was initially motivated by the animal-borne nodes in our envisioned application. We expected to draw the bulk of our considerations from stationary-only experiments. Instead, mobile nodes played a much more relevant role in our study. On one hand, the stationary-only experiments did not deliver the amount of data we expected. The connectivity patterns were not known in advance, and a multi-phase deployment was not an option, as already discussed. Mobile experiments provided a data set complementing the stationary ones. On the other hand, with hindsight, the use of mobile nodes is an effective way to explore connectivity, regardless of mobility requirements. Intuitively, a broadcasting node moving through a single, well-designed path yields a wealth of information, more varied and fine-grained w.r.t. stationary-only experiments, even considering the interference introduced by the person executing the experiments. This enables a more precise "connectivity map" of the environment, that can be used for instance to guide node placement. We believe the use of mobile nodes can become an essential element of studies aimed at characterizing connectivity in WSN environments.
3. *When WSN developers are not in charge.* Our experiments were run by someone other than the WSN developers because of opportunity. There may be other reasons, e.g., the necessity to require authorizations or safety concerns related to the target deployment area. In any case, for WSN to become truly pervasive, end-users must be empowered with the ability to deploy their own system. The lessons we learned, distilled in Section 5, can be regarded as a contribution towards this goal.

2 Deployment Scenario

Location. The community-based reserve of Junin, in the Intag region of the Imbabura province in Ecuador (0°16'19.09"N; 78°39'28.92"W) is between 1,200 and 2,800 m above sea level of the North-Western slopes of the Ecuadorian Andes. Significant portions of these mountain areas are primary cloud tropical forests, almost permanently cloudy and foggy. According to the United Nation's World Conservation Center, cloud forests comprise only 2.5% of the world's tropical forests, and approximately 25% are found in the Andean region. Therefore, they are considered at the top of the list of threatened ecosystems. The climate is tropical, and the flora and fauna incredibly rich, with about 400 species of birds and 50 known mammal species (including 20 carnivores), many probably still unchecked or even unknown. The small human community of about 50 people is 20 km from the closest village, and a 7 hour dirt-road drive from the closest town. The vegetation is made by relatively scattered mature trees, constituting the canopy, and a dense undergrowth of shrubs and epiphites. During the rainy season (November-May), when we ran our experiments, it rains every day for nearly the entire day.

WSN Equipment. Our experiments used 18 TMote Sky nodes, equipped with the ChipCon 2420 IEEE 802.15.4-compliant, 2.4 GHz radio and on-board inverted-F micro-strip

Fig. 1. Packaging

Fig. 2. In the jungle with mobile nodes

omni-directional antenna. The choice of this popular platform is motivated both by our intended use of a similar platform in our own wildlife application, and to enable comparison with similar experiments in different environments reported in the literature. Alternate hardware would significantly modify the results, e.g., an external antenna would likely dramatically increase the observed connectivity. Moreover, these motes are provided with an external flash memory, enabling storage of the experiment data.

As stationary motes were intended to be attached to trees in a very humid environment, under heavy rain, we used IP65 water-proof boxes with a transparent cover, enabling the sampling of the light as requested by the biologists. Inside each box we glued a USB female connector to easily anchor and replace the node as needed. Each box also contained a battery holder with two size D batteries and desiccant bags to protect the node against humidity. The packaging is shown in Figure 1 in the same orientation as it was attached to the trees. In contrast, the mobile node was simply a TMote Sky powered by 2 AA batteries, wrapped in a plastic bag.

3 Experiment Design

The WSN was composed of 8 nodes, placed in a cross configuration, as shown in Figure 3(a). The placement was determined as part of the stationary experiments, described next. Node 0 served as the experiment coordinator, broadcasting a message indicating the start time and configuration of each experiment. All communication took place on channel 18. Since no computer was available in-field, we used the motes' LEDs to visualize the node functionality. For example, toggling the yellow LED indicated message transmission, while toggling together the other two LEDs indicated message reception. At node boot time, a visual code for the battery voltage was shown to advise for battery replacement in case of values below 2.7 V, the minimum required to write to the flash memory. To start an experiment, the biologist pressed the user button.

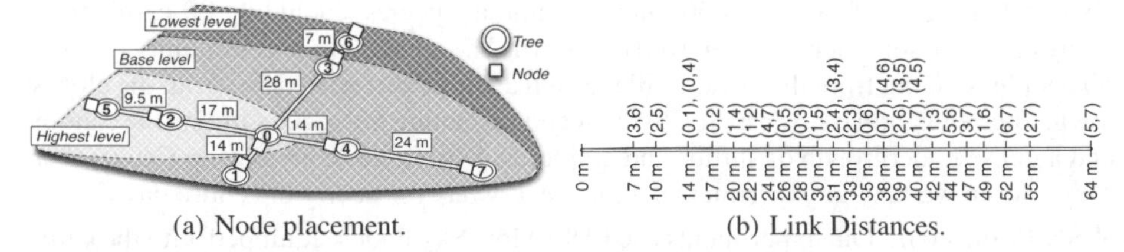

(a) Node placement. (b) Link Distances.

Fig. 3. Deployment of stationary nodes; each color corresponds to about 1 m difference

The software was built on TinyOS and without any MAC protocol, given our goal of characterizing physical connectivity. Packet collision was avoided by an appropriate transmission schedule sent at the beginning of each experiment by node 0. For each experiment, and for each link $i \rightarrow j$, we recorded in the flash the following metrics:

- *Packet Delivery Ratio* ($PDR_{i \rightarrow j}$), the number of packets received at node j over the total number of packets sent by node i;
- *Received Signal Strength Indicator* ($RSSI_{i \rightarrow j}$), the signal strength of the packets transmitted by i, as observed by the radio of j;
- *Link Quality Indicator* ($LQI_{i \rightarrow j}$), the correlation index between the symbol received at j, sent by i, and the one to which it is mapped after radio soft decoding.

3.1 Preliminary Tests

Goals. Given the lack of reported experiences in scenarios similar to ours, the primary goal of these tests was to determine the communication range, to properly place nodes in the next experiments. These experiments also investigated different power transmission levels as well as the impact of direct tree obstruction.

Implementation. The experiments exploited only node 0 and 3 in Figure 3(a). We implemented two experiments, one to determine the range of communication, and the other to investigate the effect of signal power and tree obstruction. In the former, each node sent 600 messages with an inter message interval (IMI) of 2 s. All messages were sent with -1 dBm transmission power. The LED visual feedback was used to guide the identification of the maximal communication range. In the latter experiment, each node sent a sequence of 3000 messages with a 2 s IMI, interleaving sending between the involved nodes. These messages are logically divided into 5 tests of 600 messages each, 3 at -1 dBm, commonly used in WSN deployments, and 2 at -8 dBm, to investigate the effect of reduced power. For each 600-message set we stored the aggregated average RSSI ($\overline{RSSI}$), average LQI ($\overline{LQI}$) and PDR values over all received messages.

Deployment. In all experiments node 0 was attached to a tree at 1 m height, while node 3 was placed on a chair. In the first experiment, the two nodes were in line of sight (LoS) and the biologist gradually moved the chair away from the tree while monitoring the LEDs for determining a safe communication range, which she established at 28 m. The second experiment with different power levels was run a first time with nodes in line of sight, and then again with node 0 directly behind the tree, creating a link obstruction.

3.2 Tests with Stationary Nodes

Goals. The purpose of these tests was to investigate connectivity among nodes at different distances, over a long time interval, and at different node heights.

Implementation. These experiments used the nodes as in Figure 3(a) and, as in the preliminary tests, relied on node 0 for disseminating the start time and transmission schedule. In each experiment, each node sent 215 messages with an IMI of 8 s, resulting in an interval of 1 s between nodes adjacent in the transmission schedule. The experiments were batched and ran for an entire day, interleaving 23 experiments at -1 dBm with 22 experiments at -8 dBm. Before this batch, a 1-hour *setup* experiment (with LEDs enabled) was performed, to verify connectivity and thus node placement. At the end, each node computed and stored the overall PDR, $\overline{RSSI}$, and $\overline{LQI}$ w.r.t. all other nodes.

Deployment. Node 0 and 3 were left in place after the preliminary tests. During the *setup* experiment, all the others were moved one by one away from node 0 in small steps. Based on high-level instructions, the LEDs blinking, and the communication range of 28 m determined in the preliminary tests, the biologists determined the final placement shown in Figure 3(a), yielding the set of distances covered as shown in Figure 3(b). The experiments were executed twice for a total of 2 days.

Our original idea was to deploy the nodes in a flat area, placing them first at ground level, then at 1 m from the ground, and finally at various, possibly higher heights. The rationale was to determine node placement in the least favorable connectivity conditions, close to the ground. Unfortunately, due to the delayed arrival on site (caused by lost luggage), the biologists decided to eliminate the first experiment. Moreover, due to the available terrain, highly irregular and on a sort of hill as shown in Figure 3(a), the second and third deployments were reversed. Therefore, the deployment was setup in the connectivity conditions *most* favorable, which affected the subsequent experiments. Indeed, undergrowth interfered significantly during the second test, making its results unusable. Also, node 2 failed to start some tests and its data has been excluded.

3.3 Tests with Stationary and Mobile Nodes

Goals. These experiments were initially motivated by our wildlife application, combining fixed and animal-borne nodes. When interpreting the results, however, we realized the importance of these tests in enabling exploration of connectivity at many more distances w.r.t. the static deployment, yielding more spatial continuity to data points.

Implementation. In these experiments, node 0 was carried by the biologist, who moved throughout the deployment area. Stationary nodes only listened, while node 0 broadcast messages at −1 dBm for 15 min, with an IMI of 500 ms, yielding 1,800 messages per experiment. Unlike stationary experiments, which recorded only one aggregate value for each link, in the mobile tests statistics about each individual message were recorded. This allowed us to treat each message separately, by considering the distance between the mobile node and each stationary node at the moment it was sent. Offline data correlation across nodes was enabled by timestamping the message at the sender, and saving this along with the RSSI and LQI values at the receiver. During experiments the biologist moved freely, her path recorded by a video camera carried by a second team member (Figure 2), allowing us to visualize the movements and correlate the timings.

Deployment. The placement of stationary nodes was the same as in Section 3.2, but the nodes were physically replaced as their (pre-loaded) software was different. The nodes were placed at 1 m from the ground. The mobile node was either held in the biologist hands (as in Figure 2) with the antenna parallel to her shoulders and the board facing the sky or carried chest height inside a pouch, unfortunately with undefined orientation. First, the biologist stood near a stationary node (node 2) and made simple movements of approximately 1 m amplitude along the horizontal plane at the node height and along the tree, approaching the node from four directions—front, back, right, and left. Then, the biologist moved back and forth between node 1 and 3, then between 2 and 5. Although these experiments focused on movement between a subset of the available nodes, all nodes in the network recorded message reception, thus we gathered a large amount of data. Finally, the biologist composed a path visiting all stationary nodes. Each path was repeated 4 times. In total, these experiments produced 116,448 data points. We excluded the data collected by node 7 as we verified that its short-range reception was abnormal.

Table 1. Results from the preliminary tests

Link	TX power	PDR		$\overline{RSSI}$		$\overline{LQI}$	
		LoS	Tree	LoS	Tree	LoS	Tree
$0 \rightarrow 3$	-1 dBm	86.7%	79.5%	-87 dBm	-91 dBm	99	90
$3 \rightarrow 0$	-1 dBm	84.4%	69.7%	-88 dBm	-92 dBm	98	88
$0 \rightarrow 3$	-8 dBm	24.2%	1.3%	-92 dBm	-93 dBm	80	77
$3 \rightarrow 0$	-8 dBm	11.8%	0.5%	-92 dBm	-94 dBm	77	75

4 A Mote's Life in the Jungle

4.1 Preliminary Tests

The results of the tests on transmission power and tree influence are shown in Table 1. As discussed in Section 3.1, these involved only node 0 and 3. At -1 dBm, both PDR and $\overline{LQI}$ are high. This is expected as these results are at the distance of 28 m the biologists chose as the border of good connectivity. Interestingly, our initial guess for a safe communication distance was much lower, around 10-15 m, given the presence of thick vegetation and high humidity. $\overline{RSSI}$ is low but, given the absence of radio interference in the forest, it does not significantly affect PDR. The presence of a tree right in front of a node may cause link asymmetries. With nodes in line of sight, the PDR difference between the two link directions is only 2%, but with the tree in between this increases to 10%, indicating a weaker link when communication originates near the tree. $\overline{RSSI}$ and $\overline{LQI}$ do not show marked asymmetries, although they decrease when the tree obstructs the link. With lower transmission power, PDR is non-negligible but more heavily influenced by the tree. The low $\overline{LQI}$ is consistent with the next experiments showing that 28 m is well outside the good-connectivity range at -8 dBm.

4.2 Tests with Stationary Nodes

Long-Distance, High-Quality Links. We expected the dense jungle foliage to significantly limit communication. Instead, Figure 4(a) shows that communication is almost perfect up to 20 m, although the high PDR at 19.8 m occurs with a relatively low signal strength (Figure 4(b)). Further, although the 38 m link falls well beyond the region with perfect communication, analysis over time (Figure 5) shows that this link was also perfect for more than half of the experiment duration. While this is clearly an anomaly of the setup, it clearly demonstrates that connectivity in the jungle is much different than expected. At -8 dBm, the area with perfect links is only slightly reduced to 14 m.

Fluctuations and Asymmetries of Mid-Range Links. Figure 4(a) and 4(b)) show that links with mid-range distances of 20–40 m have highly-variable quality and low RSSI. The PDR large standard deviation is best viewed over time in Figure 5, where each

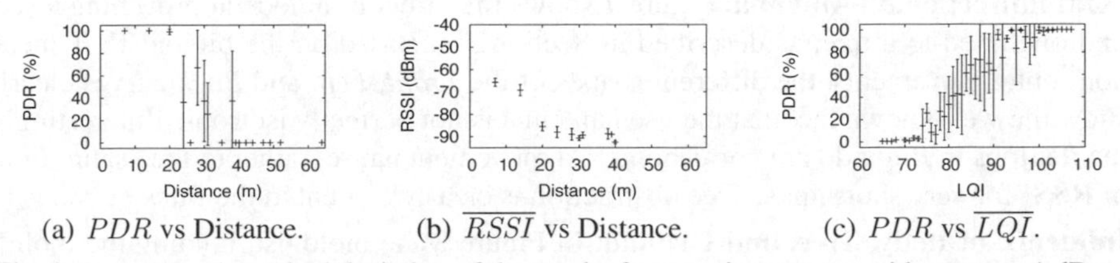

(a) PDR vs Distance. (b) $\overline{RSSI}$ vs Distance. (c) PDR vs $\overline{LQI}$.

Fig. 4. Average and standard deviation of the results from stationary tests with power -1 dBm

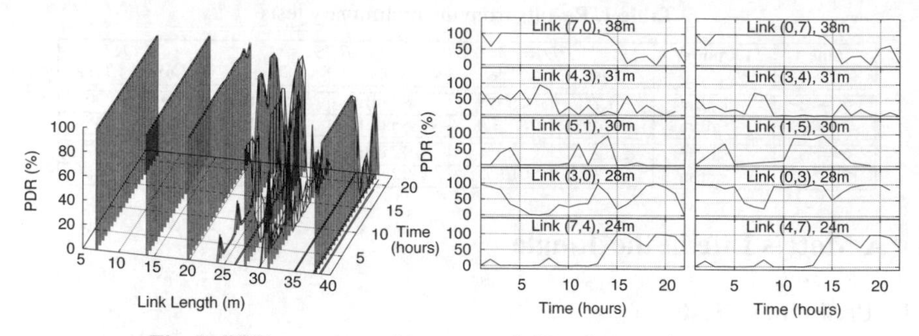

Fig. 5. PDR over time with power -1 dBm from stationary tests

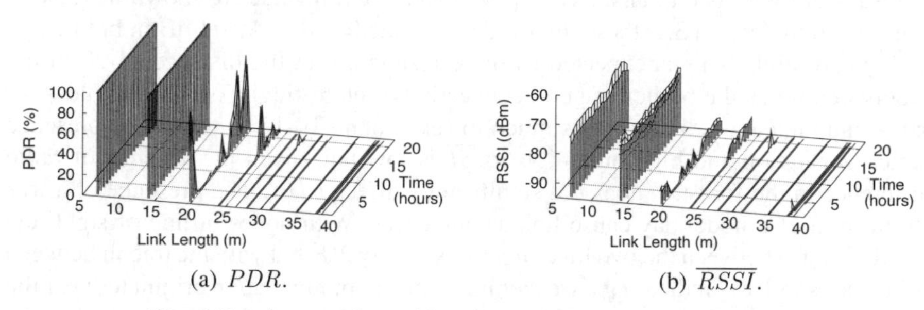

(a) PDR. (b) $\overline{RSSI}$.

Fig. 6. Results over time with power -8 dBm from stationary tests

point describes the result of one 30-min experiment for a given link. From the detail on the right-hand side of the figure, one can see that the variability is unpredictable. For example, around hour 15 some links improve while others decline. Further, some links such as $(3, 0)$ show transient asymmetries. Weather could be the culprit, and indeed it rained during the majority of these tests. Although one would expect a global decay of link quality, it is possible that humidity, rain, and pools of collected water affect communication in local, unpredictable ways, although we do not have direct observations confirming this. In any case, mid-range links clearly cannot guarantee connectivity, but they can certainly be exploited transiently by adaptive routing algorithms.

Long-Range Interference with Reduced Power. At -8 dBm, links outside the perfect communication range disappear for long periods of time (Figure 6). While these links are basically unusable, they can cause long-range interference. For example, Figure 6(b) shows messages received with very low RSSI even at 40 m. Although these distant transmissions rarely succeed, they could easily disrupt overlapping shorter-range ones.

4.3 Tests with Stationary and Mobile Nodes

"Omnidirectional" Antenna. Figure 7 shows the effect of a node approaching a second one fixed to a tree, as described in Section 3.3. Based on the biologist's 1-meter horizontal movements, the different shapes of the *Front*, *Left*, and *Back* curves clearly show the well-known fact that the used antenna is not perfectly isotropic. Interestingly, the flat tops in *Right* do not correspond to a movement pause, rather to the "saturation" of RSSI for very short links. Tree obstruction is clearly evident in the *Back* curve.

Influence of Body, Tree, and Ground. In Figure 8 the biologist, holding the mobile node in front of her chest, looped four times around nodes 1 and 3. We decomposed the

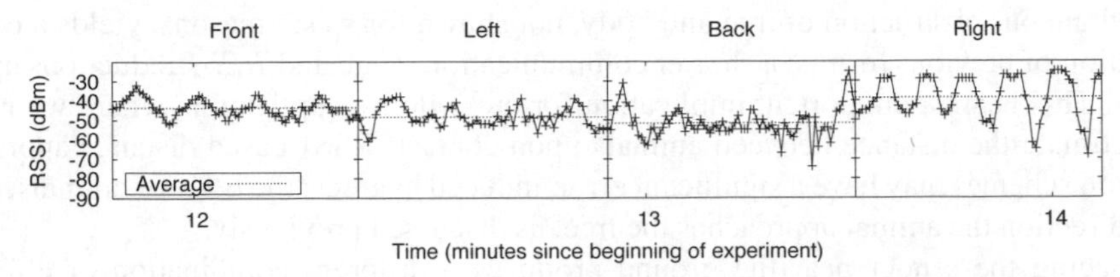

Fig. 7. Node 0 approaching node 2, attached to a tree, from different directions

data trace to distinguish the possible obstructions. For example, when walking from 1 to 3, the tree obstructed communication received at 3 (Figure 8(b)), and the body obstructed receptions at 1 (Figure 8(c)). As a reference, we chose the line-of-sight case: reception at 1 when walking from 3 to 1 (Figure 8(a)). The same experiment was run with the mobile node held a few centimeters from the ground (Figure 8(d)).

Trees induce a reduction up to 20% on $\overline{RSSI}$ in short links (< 20 m), while longer links are not affected. The body also reduces $\overline{RSSI}$ in short links, but more significantly, up to 40%. Moreover, the body reduces the maximum communication range by 10 m, as denoted in Figure 8(c) by a nearly-zero PDR beyond 30 m. As expected, the

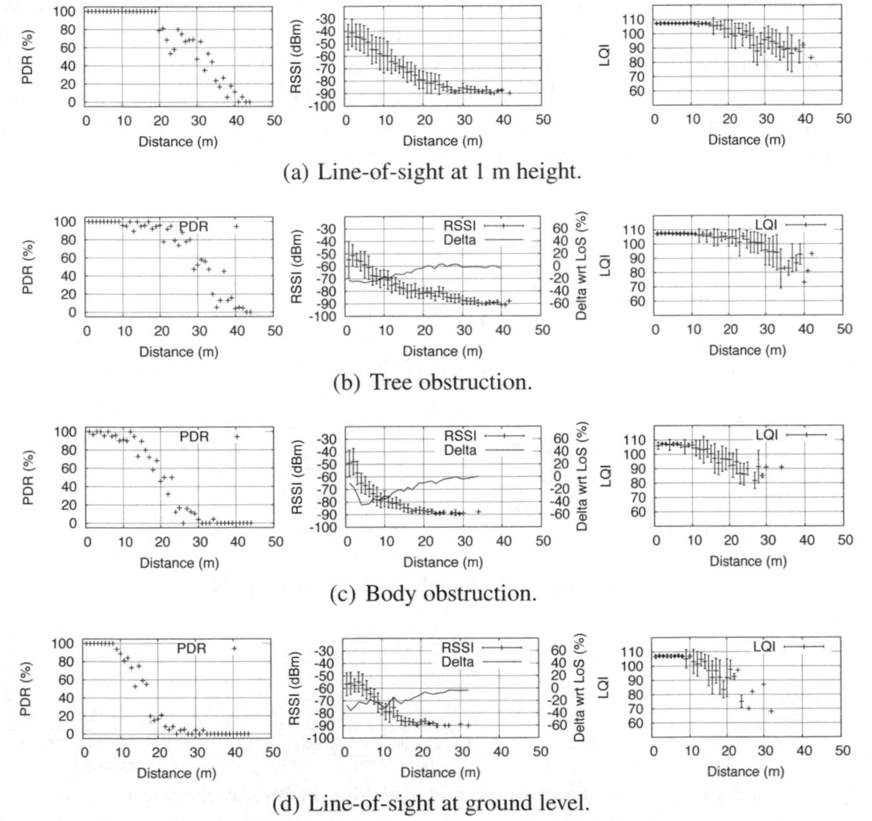

(a) Line-of-sight at 1 m height.

(b) Tree obstruction.

(c) Body obstruction.

(d) Line-of-sight at ground level.

Fig. 8. Effect of tree, body, and ground on communication. The line in the RSSI plots shows the delta in percent w.r.t. the line-of-sight shown in (a).

simultaneous obstruction of tree and body, not shown for space reasons, yields a combination of previous results: a shorter communication range and $\overline{RSSI}$ reductions up to 60%. This bears an important implication for our wildlife application, where we need to estimate the distance between animals upon contact: RSSI-based distance approximation schemes may have a significant error, induced by trees, the body of animals, and the direction the animal approaches the tree, as discussed previously.

Placing the sender near the ground produces a different combination of effects. Specifically, the line-of-sight communication range is much shorter than in Figure 8(a), but the $\overline{RSSI}$ is affected by at most 20%. As this scenario is the closest to our target deployment with tagged animals, it warrants additional study.

4.4 An Evaluation of Mobile Nodes as Connectivity Probes

We take a step back from the data analysis to consider our data collection methodology, specifically, comparing the results of stationary test against those with mobile ones.

Aggregated Mobile Tests vs. Stationary Tests. Thus far we have looked only at excerpts of the mobile traces, extracting cases with specific characteristics. Here, we aggregate all data points collected over all node movements, with the results shown in Figure 9(a)–9(c). To plot PDR, we calculate the distance between the mobile and each stationary node, then plot the number of messages received over those sent at each distance. $\overline{RSSI}$ and $\overline{LQI}$ are instead shown as the average and standard deviation over all the messages received along links of a specific length. We then compare these data to those collected in the stationary tests of Figure 4, by plotting the percentage difference in Figure 9(d), only for the points studied in the stationary scenario.

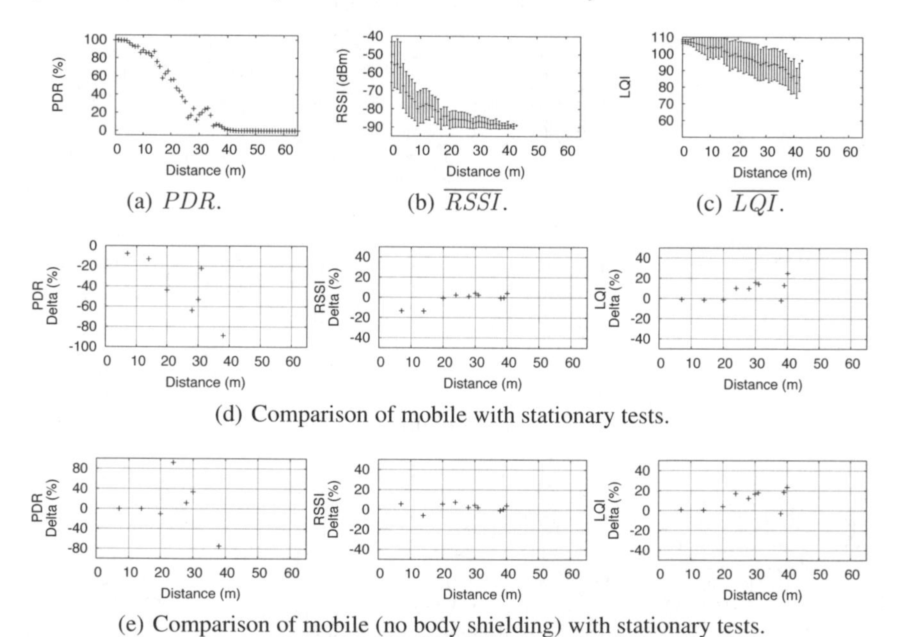

Fig. 9. Aggregated results over all 11 mobile experiments. In (d) and (e), the difference in PDR for the links longer than 38 m is outside of the chart range.

In the mobile scenario, the reduction of $\overline{RSSI}$ on short links (< 15 m) is likely attributable to body interference as observed in Figure 8(c). From the PDR comparison in Figure 9(a), we note that at all distances, the mobile scenario produces *worse* results, meaning that the PDR at a given distance is lower in the mobile scenario than in the stationary. To understand the implications, consider that we intend to use the results of this study to plan a future deployment. If we base this deployment only on the results of the mobile study, all stationary nodes in our future deployment would certainly be connected. Instead, if we base our fixed node placement on the stationary results, we would erroneously expect to communicate with mobile nodes carried by animals at the same distance. In other words, the mobile case underestimates the communication potential of stationary nodes while the stationary overestimates communication to mobile nodes.

Interestingly, Figure 9(c) shows better quality links in the mobile scenario. While this is opposite from the observations of PDR, the stationary experiments showed that LQI varied significantly throughout the day. Instead, the mobile experiments were concentrated in less time, and may have taken place in favorable connectivity conditions.

Figure 9(e) accounts only for the data recorded in conditions similar to those of the stationary only tests, i.e. removing the body shielding and using the data from Figures 8(a) and 8(b), namely LoS and tree-only obstruction. For short links (< 20 m), values are in agreement while longer links are hampered by interference from the ground and dense low-level foliage in the mobile scenario. In the stationary tests, nodes were always within LoS, therefore the undergrowth had minimal effect.

Statistical Relevance of Mobile Tests. The experiments run with a mobile node made it possible to explore the physical space in a continuous fashion, spreading the collected data points over more distances w.r.t. stationary-only tests. To understand the effectiveness of this approach, Figure 10 compares the average number of messages received in 1 hour for each distance covered in the mobile case, to the number of messages that the stationary experiment would receive with the same IMI as the mobile nodes, i.e. 500 ms. Recall that to avoid collisions, our stationary experiments used a 1 s IMI. The distribution of the tested distances is naturally biased by the executed movements. Nonetheless, even without a guided motion plan, all distances less than 40 m have been tested by at least 400 messages, i.e., 25% of the messages sent by the stationary tests for each link. The ability of the mobile node to cover so many distances clearly motivates its use as a probe to characterize connectivity.

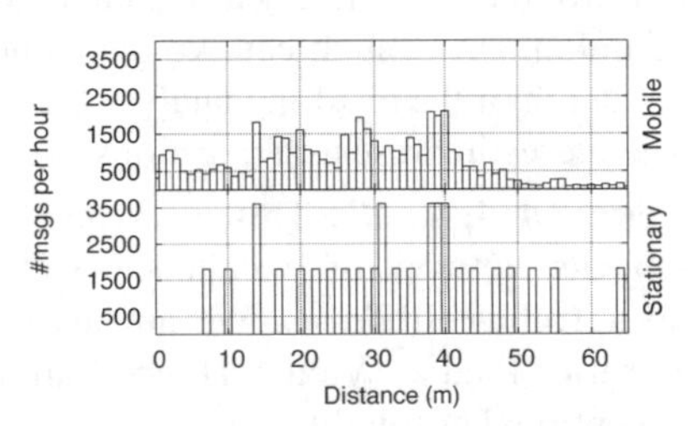

Fig. 10. Number of messages sent per hour along links of a given distance by stationary and mobile tests

5 Lessons Learned and Future Work

Our experiments were run in a challenging scenario by biologists without WSN expertise, with limited equipment, and in isolation. We have never faced this combination in our previous real world deployments, and we learned interesting lessons.

Mobile Nodes: Application Insights or Connectivity Probes? It was the biologists who requested experiments with mobile nodes, to concretely understand what WSNs could offer them. Nevertheless, we learned that the use of mobile nodes, despite the inherent imprecision, is useful for characterizing an unknown environment and guiding the actual deployment. Further work is needed to explore the opportunities of this technique and understand its limitations, e.g., the difficulty to capture long-term variations.

The Role of LEDs. In our study, the node output had to be simple yet informative enough to guide the biologists. Our solution, based on giving a visual clue only about send/receive operations, contributed to the creation of very long links between stationary nodes which in turn contributed to the failure of the second set of stationary experiments, as mentioned in Section 4.2. A visual representation of the RSSI values (e.g., represented by a "histogram" using the three LEDs), would have led to shorter links, which would have produced meaningful data even in the second set of experiments.

Testing Blindly. Our experimental campaign involved many decisions taken blindly. We did not have an understanding of the environment based on previous studies. We did not have a well-defined methodology for performing this kind of experiments, and none yet exists in the WSN field. Finally, we could not modify experiments based on intermediate results. We partially reduced the unknowns by breaking down experiments into phases with well-defined outputs. Examples are the preliminary tests (Section 3.1) and the 1-hour *setup* phase preceding the stationary tests (Section 3.2). These enabled the biologists to take informed decisions autonomously, partially obviating the absence of WSN experts in-field. Nevertheless, this did not avoid incorrect decisions, and could not provide answers for unanticipated questions (e.g., the cause of high time variance of links). How much can we reconcile the autonomous execution of experiments and the depth of the resulting analysis? To what extent can we automate the process? These are interesting research questions and the subject of our ongoing work.

Acknowledgments. This work has been partially supported by the EU Cooperating Objects Network of Excellence (CONET: FP7-2007-2-224053) and by Martin Stanley (Holly Hill Trust). We are also indebted to the Junin community, in particular Rosario Piedra, Victor Hugo Ramirez, Olga Cultid, for logistical support, and exceptional hospitality.

References

1. Miluzzo, E., Zheng, X., Fodor, K., Campbell, A.T.: Radio characterization of 802.15.4 and its impact on the design of mobile sensor networks. In: Verdone, R. (ed.) EWSN 2008. LNCS, vol. 4913, pp. 171–188. Springer, Heidelberg (2008)
2. Mottola, L., Picco, G., Ceriotti, M., Guna, S., Murphy, A.: Not All Wireless Sensor Networks Are Created Equal: A Comparative Study on Tunnels. ACM Transactions on Sensor Networks (TOSN) 7(2) (2010)
3. Polastre, J., Szewczyk, R., Culler, D.: Telos: Enabling ultra-low power wireless research. In: Proceedings of the 5th International Conference on Information Processing in Sensor Networks, IPSN (2005)
4. Srinivasan, K., Dutta, P., Tavakoli, A., Levis, P.: An empirical study of low-power wireless. ACM Transactions on Sensor Networks (TOSN) 6(2), 1–49 (2010)
5. Tolle, G., Polastre, J., Szewczyk, R., Culler, D., Turner, N., Tu, K., Burgess, S., Dawson, T., Buonadonna, P., Gay, D., Hong, W.: A macroscope in the redwoods. In: Proceedings of the 3rd International Conference on Embedded Networked Sensor Systems, SenSys (2005)

Deploying Wireless Sensor Networking Technology in a Rescue Team Context

Ben McCarthy[1], Socrates Varakliotis[2], Christopher Edwards[1], and Utz Roedig[1]

[1] Computing Department, Infolab 21, Lancaster University,
London, WC1E 6BT, UK
{b.mccarthy,ce,u.roedig}@comp.lancs.ac.uk
[2] Computer Science Department, University College London,
Lancaster, LA1 4WA, UK
s.varakliotis@cs.ucl.ac.uk

Abstract. The computing department at Lancaster University are currently involved in the ongoing deployment of an advanced communications system designed to support the requirements of search and rescue teams. This system is based around the concept of using an all IP infrastructure to provide multi-functional data communications (such as group voice calling, live video streaming and location updates) to highly mobile vehicles and personnel in challenging environments. In addition to these types of data communications there is also a requirement to reliably transmit different types of sensor data information from the individual rescue team members, their vehicles and the casualties they locate and rescue. In this paper we describe the work we have carried out to incorporate an IP based sensor networking approach into our existing communications system deployment that we have in place with the Morecambe Bay Search and Rescue Team, in order to support Mobile Sensor Networks. In addition, we present results from our experimentation with our deployment that is specifically focused on the issue of wireless interference that our Mobile Sensor Networking solution is potentially subjected to.

1 Introduction

In general, search and rescue services have an inherent reliance on their communications solutions because their success is often closely related to their ability to accurately exchange information about a rescue situation between their team members. However, most search and rescue teams are still vastly under provisioned with regards to their communications equipment, with most teams still relying purely on push-to-talk radio solutions. Lancaster University are currently involved in an on-going effort to develop a powerful new communications solution that leverages Internet technologies to provide search and rescue teams with advanced functionality such as real-time localisation and mapping, real-time video streaming from vehicles and individuals, advanced voice services (i.e. tailored group calling, rather than indeterminate broadcasting of voice calls) and

P.J. Marron et al. (Eds.): REALWSN 2010, LNCS 6511, pp. 37–48, 2010.

real-time delivery of sensor data from the field of operation. In this paper we specifically focus on the experiences we have gained from incorporating an all IP sensor networking solution into our current mobile communications deployment with the Morecambe Bay Search and Rescue team (UK) [1]. In particular we provide an overview of the design of our solution as a whole and its reliability for delivering timely health statistics in challenging environments about individual rescue team members and the casualties they locate, as well as environmental data recorded by rescue team vehicles.

Our efforts with the Morecambe Bay Search and Rescue team demonstrate a real world deployment of Mobile Sensor Networking (MSN) technology and illustrate how sensor networks deployed in this context may often have to be considered as a component of a bigger overall system that has the potential to cause interference to their ability to deliver timely sensor data. From our deployment experience we have encountered scenarios where radio communication in over-lapping frequencies has caused sensor data delivery to be affected and in a critical scenario such as search and rescue the loss of reliable health statistics about a specific team member could be extremely damaging. The issue of radio interference arises because of our system's joint reliance on multiple wireless communication technologies that each transmit in the 2.4GHz ISM band of radio frequencies, namely the 802.11g, Bluetooth and 802.15.4 protocols. To investigate this problem further we constructed a number of tests designed to identify the real effects on sensor data delivery that are experienced with our system when used in the rescue team's operational environment.

The rest of this paper is presented as follows: In Section 2 we detail the nature of the rescue services deployment we are currently involved with and highlight how our system architecture maps onto their operational model. In Section 3 we focus on the way we have provisioned for sensor networking in the mobile context the rescue team operate in. In Section 4 we outline the specific hardware we used, as well as present the in-field interference experimentation we carried out. In Section 5 we provide an overview of related work that has been carried out in this field. Finally, in Section 6 we conclude with a discussion about our findings and overall experience of incorporating sensor networking into our search and rescue system deployment.

2 Trial Deployment

The Morecambe Bay Search and Rescue Team are a non-profit organisation that operate search and rescue missions primarily on and around the Morecambe Bay area in the North-West of the UK. In total their team consists of 16 operational members, all of whom are voluntary workers, devoting their time and effort for free. The Morecambe Bay area is the largest expanse of inter-tidal mudflats and sand in the United Kingdom and in total covers an area of 310 km^2. At low tide the sea retreats leaving a massive expanse of open sand that people, vehicles and animals use to walk/travel on and even work on (picking shellfish). In this state the bay area is deceptive and appears safe to access, but in reality it is peppered with treacherous pockets of quicksand and hidden channels that quickly fill areas with sea water when the tide returns. Due to these conditions many unsuspecting people (sometimes with vehicles) and animals get trapped, at which point their lives are immediately in extreme danger. For this reason the Morecambe Bay Search and Rescue team was created, to be able to cope with the specific requirements of rescuing people in the bay area's adverse and dangerous environment. More recently, the Morecambe Bay Search and Rescue team now also provide official support services to the area's fire brigade in land based situations which require their specialist expertise and equipment.

2.1 Rescue System Deployment

The rescue team currently have four primary vehicles that they use in their rescue operations (shown in Figure 1): 2 Hagglund BV206 All Terrain Emergency Rescue Vehicles, 1 Landrover 4x4 ambulance and 1 high-speed airboat. These vehicles carry out search operations (in incidents where a casualty's location is not already known), transport the rescue team members to an incident area and also act as a mobile command post for operation coordination. The communications system we have deployed is based around the notion of interconnecting wireless Vehicle Area Networks projected in and around each of the rescue team vehicles with wireless Personal Area Networks projected by each of the individual team members. Into each vehicle we have fitted a ruggedized, bespoke, vehicle specification Mobile Router (MR) which is powered from the vehicle's main power supply but that also trickle-charges a dedicated battery pack for use when the vehicle's own power supply is cut off. Each vehicle mounted MR has a number of wireless interfaces (each fitted with a dedicated external roof mounted antenna) including 2 802.11b/g interfaces (1 for interconnecting with other MRs, 1 for projecting a connectivity hotspot around the vehicle) and a cellular data modem capable of connecting to GPRS, EDGE and HSDPA services. In addition, we have also fitted the airboat vehicle with an INMARSAT BGAN vehicle terminal which is capable of maintaining a consistent connection to INMARSAT's broadband data service while the vehicle is in motion.

Fig. 1. Morecambe Bay Search and Rescue Team Vehicles

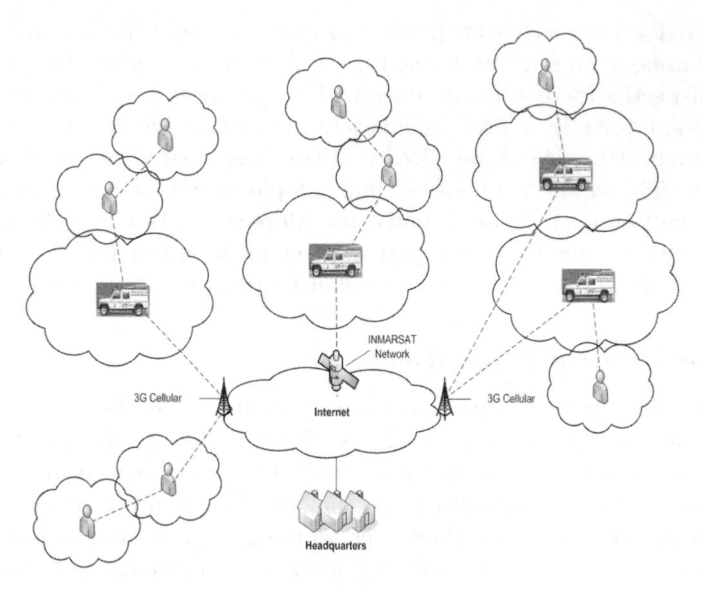

Fig. 2. Communications Model

In our model, as well as each vehicle containing a MR, each individual team member also carries a similar device designed in a suitable form factor for personal use. In particular, apart from being physically smaller than the vehicle mounted MRs the personal MRs also have a completely self-contained rechargeable Li-ion power supply and an additional 802.15.4 wireless interface. Once operational all of the MRs attempt to establish their own direct connection to the Internet via their cellular interface whilst simultaneously forming connections with other MRs via 802.11g. This process results in an interconnected Mobile Ad hoc Network (MANET) like the one illustrated in Figure 2 where any nodes capable of communicating directly with each other do so via multi-hop 802.11 and all other communication is performed using the Internet as a backbone. Using the Unified MANEMO Architecture (UMA) routing approach (discussed in further detail in the following section) allows us to provide unchanging, globally reachable IPv6 addresses to every node in the network, i.e. both the MRs and any host devices that subsequently connected to them, irrespective of how the topology of the network changes. This in turn allows data to be transmitted between every member of the rescue team, whether they are located at the headquarters, in one of the vehicles or out in the field, permitting communication such as group voice calls, streaming video delivery, location plotting and real-time sensor data monitoring. In the following section we focus in on how we have achieved this real-time sensor data delivery by incorporating Wireless Sensor Networking (WSN) technologies into our system to produce a fully functioning Mobile Sensor Networking (MSN) deployment.

3 Rescue System Sensor Networking

The term "Mobile Sensor Network" (MSN) is loosely used to describe WSN solutions which are capable of moving freely, potentially utilising different points of attachment to the Internet to continuously transmit real-time data about an entity or its surrounding environment as it changes location. This type of network of sensors is therefore ideally suited to gathering a broad range of data about a specific mobile entity, such as a vehicle or person, allowing their status to be continually monitored irrespective of their movement. In essence what we have achieved by incorporating our sensor networking solution into our search and rescue team deployment is a fully integrated real-world example of vehicle and person based MSNs. From a communications perspective, our approach achieves this by integrating the use of two key network layer technologies. Firstly the Unified MANEMO Architecture (UMA) [2] is used to provide the inter-communication capabilities of the rescue system, and therefore ensures that any IPv6 communication can be transmitted to any remote data sink in the Internet. Secondly, the 6LoWPAN adaptation protocol [3] is used to permit the efficient transfer of IPv6 packets to and from the resource constrained sensor nodes that are connected to the vehicle/person area networks via their low power 802.15.4 links.

The common features of WSNs are low bandwidth, constrained memory and limited computational power. Initially manufacturers introduced proprietary protocols to drive WSNs with customised link-layer solutions, assuming that IP was too resource-intensive to be scaled down to operate on the micro-controllers and low-power wireless links used in WSN settings. The 6LoWPAN protocol has addressed this situation and is what we use to provide the individual sensor nodes with IPv6 connectivity[3]. With 6LoWPAN, packet transfer from a sensor node to the network via a gateway is achieved by first fragmenting large IPv6 packets into chunks of 127 bytes or less. Once all fragments reach the gateway, packet re-assembly takes place and the composed IPv6 packet is subsequently routed to the Internet. The most commonly used header fields of the original IP packet may also be compressed as they are not required for routing within the sensor network, if layer 2 meshing is used. This compression and header stacking along with cross-layer optimisations result in low overheads, which translate to efficient transmission of IPv6 datagrams over low power networks. The overall savings can reduce the complete standard IPv6 packet (40 byte headers) down to an optimised few bytes only (around 2 bytes, at best, in typical uses) for Wireless Sensor Networks.

Fundamentally, UMA is a technique designed to enable mobile networks to perform persistent, uninterrupted IPv6 communication over the Internet, regardless of their potentially changing location and Internet access connection. In addition, UMA also ensures that mobile networks of devices can interconnect and communicate directly or share their Internet connections with other networks that cannot obtain their own Internet access connection, thus proliferating the availability of Internet access over a greater area. UMA achieves this by employing a technique that leverages the global connectivity characteristics of the NEtwork MObility Basic Support (NEMO BS) protocol [4] with the localised multihop communication support provided by MANET protocols. With UMA, every Mobile Router is registered with a corresponding Home Agent (HA) that records the its changing point of attachment to the Internet. This HA is located in the home network of the Mobile Router (i.e. the location it originates from, such as rescue team's headquarters) and intercepts packets destined for the mobile network whenever it is not directly attached to the home network. As the Mobile Router moves and potentially roams across different access networks or in-directly utilises another Mobile Router's Internet connection, it updates the HA with its new attachment point to the Internet. The HA then forwards all packets destined for the Mobile Router's network (either directly to it or in-directly via any other Mobile Router that is providing it with an Internet connection) via a bi-directional tunnel. This approach therefore keeps the mobility of the network transparent to any nodes it communicates with other than the HA and also prevents the traffic it generates from being Ingress Filtered in the access networks it visits. In total, this ensures that packets sent to and from the mobile network can use the same persistent IP address range to communicate regardless of its underlying mobility, and provides a highly robust solution because redundant, heterogeneous links to the Internet can be established and utilised if/when existing links fail.

4 Hardware/Software Setup and Experimentation

In one of our previous studies [5] we presented results from our initial lab based experimentation that acted as a proof of concept for our MSN solution and demonstrated the early potential of our approach. In this paper we confirmed the successful implementation of our solution using a Lippert Embedded Systems [6] CoolMoteMaster device, augmented to operate as a UMA Mobile Router also. This device however was not suitable for actual deployment in our scenario due to its form factor (too bulky and power hungry for personal use) and its non-ruggedized design. So instead we set about incorporating the concept we had proven in a lab environment into our existing ruggedized Mobile Router devices that we had already designed for real world deployment. To achieve this we needed to successfully incorporate an 802.15.4 interface into our Mobile Router design that could also perform the role of a 6LoWPAN gateway node and handle the packets sent too and from the sensor nodes appropriately. Both the gateway node and the sensor nodes are based on the popular Tmote Sky device (Telos Rev.B hardware platform, or 'mote'), the gateway node is directly attached to the Mobile Router board and powered via USB, whereas the sensor nodes are powered by battery. In each case we run the Contiki open-source operating system [7] which features the uIP network stack for IP communication with the motes. The uIP stack on the sensor nodes was further extended with a *sensor-side implementation* of the 6LoWPAN adaptation layer. On each sensor node there is a running measurement software component that upon initialisation performs basic IPv6 address auto-configuration by using a predefined network prefix known to the gateway. The measurement component then records information from the appropriate sensors, for the vehicle it records environment data

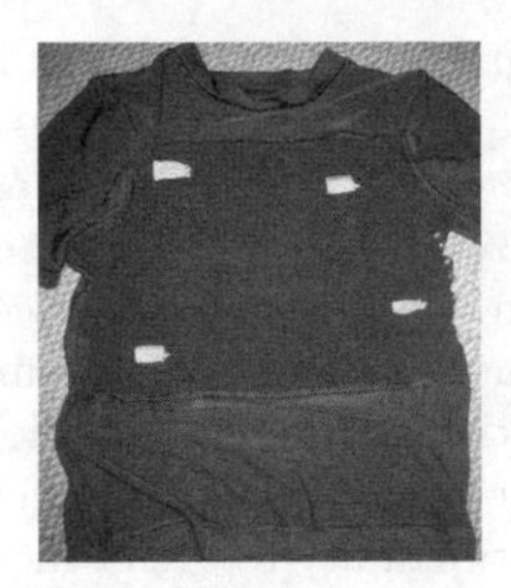

Fig. 3. SmartLife Technology "HealthVest"

including light, humidity and temperature measurements from the Tmote's embedded sensors. For the rescue team members we have interfaced the Tmote board to a prototype intelligent garment made by SmartLife Technology [8]. Their garment, known as the "HealthVest" (shown turned inside out in Figure 3) is a close fitting undergarment that incorporates their patented woven sensor technology to continuously monitor heart rate and electrocardiography (ECG) information.

In both cases, once this data is captured it is then placed in IPv6 packet payloads and transported with UDP to our remote sink application (a remote server located at the headquarters which provides the end user with combined mapping and personnel/vehicle localisation and sensor monitoring) in the following way. First, IPv6 packets larger than 127 bytes are appropriately fragmented by the 6LoWPAN sensor-side module on the source node for transport within the sensor network. The global IPv6 addresses in the headers are compressed as the sensor network only uses 16-bit link-layer identifiers. Once all related packet fragments reach the gateway, the corresponding gateway-side 6LoWPAN module re-assembles them into a full IPv6 packet and adds the decompressed destination IPv6 address of the target sink. The complete IPv6 packet is then passed to the kernel for further processing (i.e. IP routing using UMA, towards the data sink).

One final hardware device that is significant to our deployment and the experimentation presented in this paper is the Nonin Onyx II 9560 pulse oximeter [9]. This device is the world's first wireless enabled fingertip pulse oximeter and it works by periodically transmitting heart rate and blood oxygen level information over a paired Bluetooth connection. The ability to remotely monitor pulse oximetery data related to a casualty was a requirement specifically requested by the Morecambe Bay Search and Rescue team. This device offered us the perfect solution to this requirement but unfortunately operates over a different wireless technology to the others we already support and therefore its integration into the rescue system introduced further considerations about the wireless spectrum availability.

4.1 In-Field Experimentation

In total, at any one time in our rescue team communications system there can be up to 5 different wireless interface types operating at once. Whilst the satellite and GSM wide-area backhaul connections that we use in the system operate in distinct frequency ranges that do not cause interference with any of the other radios (INMARSAT BGAN receivers transmit and receive between 1525.0–1559.0 MHz and 1626.5–1660.5 MHz, O2 UK 3G service operates between 2125–2135 MHz), the 802.11g, Bluetooth and 802.15.4 interfaces all operate in the 2.4 GHz ISM band of frequencies. Each of these radio technologies transmitting in the 2.4GHz band have the potential to create contention with each other for access to the radio medium and therefore cause communications disruptions. This fact is further exacerbated by the proximity to each other that these interfaces can be expected to naturally operate in during a typical mission. To further elaborate on this statement we must first consider the primary purpose of these interfaces and then consider the context in which they are used. For example, in a typical rescue scenario, the 802.11g interface may be used extensively to carry multiple different types of communication (voice, video and location/sensor data). At any given time it may be relied upon to transmit relatively bandwidth intensive communication streams, therefore occupying one or two channels of the 802.11 radio spectrum which in turn can equate to potential interference across around 8 of the 802.15.4 channels. This type communication will also be relatively long lived in comparison to periodic sensor data transmission, and so it is extremely likely that sensor data transmission will occur at the same time as intensive 802.11 activity. One further communications consideration in our scenario occurs when a casualty is located and is transported back to safe ground and then hospital for treatment. As described in the previous section we have provisioned for the use of wireless pulse oximeters with our system during in-field operation that permit pulse rate and blood oxygen levels to be transmitted to awaiting hospital staff. To support this we auto-establish a serial-over-Bluetooth connection between the pulse oximeter and the onboard Bluetooth interface integrated in the rescue team member's Mobile Router. This means that from the moment the pulse oximeter is placed onto the casualty's finger, a third (periodic) transmission is introduced to the 2.4GHz ISM band. What this overview of the wireless communications landscape created by our rescue system shows is that it is therefore extremely important to ensure that radio interaction in our system operates fairly and that wireless transmissions do not unacceptably affect one another to the detriment of the entire system.

In order to test the interference effects experienced by the sensor networking data transmitted in our deployment we setup a series of experiments that involved the use of a number of different sensor networks and a number of different Mobile Routers in a range of different wireless configurations. In each test we deployed 1 Vehicle Mobile Router and then between 1 to 3 Personal Mobile Routers (introducing one after another to incrementally increase the level of interference experienced), with a simple hierarchical topology where all Personal Mobile Routers connected directly to the vehicle. We then connected a wireless 802.11 webcam to each of the personal area hotspots that each team member projected from their Personal Mobile Router. We then drove traffic over both wireless interfaces of the Personal Mobile Routers by streaming live video from the webcams up to the vehicle cabin. Finally, we activated the data sink in the vehicle also (in our deployment, all vehicles and the headquarters accept and display location and sensor data to the coordinators) and then configured each specific test. We configured the devices in each of our tests to forcibly create three different levels of combined interference: High, Medium and Low. These interference levels refer to our manipulation of channel selection for each of the 802.11g interfaces and the 802.15.4 interfaces, where:

- High: 2 802.11g on channel 7, 802.15.4 gateway/nodes on channel 18.
- Medium: 2 802.11g on channel 7, 802.15.4 gateway/nodes on channel 20.
- Low: 2 802.11g on channel 7, 802.15.4 gateway/nodes on channel 26.

In keeping with the focus of this paper on real deployment experience we decided to present our findings based on the most important aspect to the end user, i.e. the data arrival rate at the data sink. In our deployment (as with any other) the rescue team coordinators are interested in whether they are receiving sensor data correctly or whether they are not. As our sensor networking approach is fully IP compatible we opted to measure packet arrival rate using the Wireshark network protocol analyser which provides the ability to filter out specific packet types, offers accurate packet arrival times and also has additional tools for overall analysis. For each test, the sensors periodically transmitted their data once every second for a duration of 1000 seconds and we repeated the motions of the test 5 times over to gain average packet loss rates. Then at the end of each test run we connected a Bluetooth pulse oximeter to one of the rescue team member's Mobile Router and began transmitting casualty health statistics over each resulting network topology to observe whether the pulse oximetry data was affected.

The percentage loss experienced at the data sink in each of the configurations we tested can be seen in Figure 4. What is immediately evident from these results is the obvious correlation between the high level of sensor data loss and the high level of radio interference present. When operating on completely none-overlapping radio frequencies the level of packet loss experienced was encouragingly low. When operating in rescue missions, team members originating from one vehicle often work together in a designated area. Each vehicle will tend to carry around 4 rescue team members, 3 that will alight at given locations and 1 driver that remains in the vehicle at most times. With this operational model it would be straightforward to devise a suitable channel separation scheme between each of the team members belonging to one vehicle that would ensure their radio communications were not in over-lapping frequencies for the bulk of their time in the field of operation. Less encouraging however were the levels of loss we recorded at the "medium" level of interference. In a planned frequency allocation model like the aforementioned channel separation scheme based on vehicle assignment, this configuration effectively represents the encroachment of radio transmissions from devices outside of the planned model, i.e. when other vehicles converge in a single location. This scenario is not atypical and must therefore be expected to occur, in which case our results show the level of successful delivery of sensor data really begins to suffer.

Interference Level:	1 Mobile Router	2 Mobile Routers	3 Mobile Routers
Low:	1.4%	1.3%	1.8%
Medium:	4.6%	16.7%	27.1%
High:	42.6%	44.5%	46.0%

Fig. 4. Percentage loss in each configuration

One interesting result was the negligible effect that adding additional radio interference to the already high level of over-lapping 802.11 transmissions had when further Mobile Routers were introduced. The radio interference effectively appeared to reach a saturation point that it never rose significantly above. One explanation for this behaviour could be the specific webcam hardware used in our deployment. At present we use Panasonic webcams (because of their native IPv6 support) these webcams stream their video captures using TCP and as a result will aggressively back off under heavy interference conditions. Away from the average loss rates entirely, we also observed that during the "High" level of interference sensor data packet loss was rarely consecutive. When a sensor data packet was lost, in most cases the following packet in the transmission (i.e. the next 1 second interval) could be expected to arrive. This was observed with noticeable consistency to the point where in tests that yielded almost 50% loss, packets could be seen to arrive in an almost a 1 on, 1 off fashion.

5 Related Work

At present we are unaware of any successful deployment attempts that have aimed to provide IP based Mobile Sensor Networking in a challenging environment such as search and rescue. However, whilst work related to our system deployment as a whole is scarce at the moment, there is a lot of existing research related to the radio interference experienced between 802.11 and 802.15.4 which is a subject we focused on in this paper. One of the closest studies to the analysis we carried out in our deployment was a recent paper specifically focusing on the interference properties experienced between two sensor nodes communicating with one another in a Body Area Network (BAN) [10]. The authors of this paper provide a very complete and thorough analysis of the effects of interference on a point-to-point 802.15.4 link when it encounters 802.11 traffic in a similar radio frequency. They are able to draw stark parallels between an increased successful transmission rate and increases in the transmit power that the sensor nodes operate at.

Additional work that is related to the particular sensor data that we transmit in our deployment has also been carried out from the perspective of medical applications in hospitals [11]. In this study the authors set out to determine the suitability of 802.15.4 for the purpose of carrying health statistics information in a typical hospital environment. However whilst the sensor data and end devices discussed in this study is related, the operational environment is naturally very different and the authors attained their results through simulation.

As well as studies analysing the impact of 802.11 on 802.15.4, there also exists studies that have considered this problem from the opposite angle of how

802.15.4 can impact on 802.11's overall capability. Interestingly there are papers with conflicting conclusions in this research space [12] [13] with studies finding no effect and another recent study finding effects on 802.11 in specific circumstances. In our deployment any effect on 802.11 by 802.15.4 was not noticeable and will unlikely ever be if we continue to operate a model of 1 second sampling times and single sensor gateway/node pairings between each individual Mobile Router. For the foreseeable future it seems that this model is sufficient for our requirements and at that node density the 802.15.4 transmissions simply are not frequent enough to cause significant disruption to any 802.11 communications.

6 Conclusion

In this paper we have provided an overview of our mobile networking deployment activities with the Morecambe Bay Search and Rescue Team. We have focused in particular on our recent work related to incorporating an innovative all-IP sensor networking technique into the existing IP infrastructure established by our rescue system communications solution. Apart from detailing the technologies and hardware/software components we have used to achieve these outcomes we have also provided an insight into some of the experiences we have gathered so far. Building on these experiences we have also outlined some in-field testing we performed to ascertain the suitability of our approach for reliably transmitting important sensor data information throughout the rescue team network. Specifically, this testing concentrated on the potential negative effects of radio interference that can be experienced in large, multi-function communication systems like the one we have deployed.

In particular, we found radio interference from intensively used 802.11g wireless interfaces did have a negative impact if channel overlap existed with the 802.15.4 spectrum that was utilised. However, what we also observed was that even in very high levels of interference there would often be little consecutive loss. Therefore when a packet of sensor data was lost, in all but the worst cases the next packet could statistically be expected to be delivered. This means that the importance of the data being carried must be taken into consideration, and more specifically, the criticality of every single reading produced. If we take our deployment for example, the significance of every single sensor data reading is debatable. Certainly from the perspective of an environmental reading from a vehicle the levels of loss experienced are unimportant, but this can even be said of the health data of the individual rescue workers. During an operation a team coordinator will monitor the progress of their team looking at location and movement, if the weather is bad or worsens they may casually maintain a watch on the overall health statistics of their members, but any data loss would most probably go unnoticed. However, if a team member was identified as going overboard into deep water in Winter this interaction could quickly change and every reading could become extremely important. Related to this is the criticality of a casualty's health statistic data, however whilst this data is seen as being extremely important the use of Bluetooth and its Adaptive Frequency Hopping Algorithm in this case suitably addresses this problem.

Finally in this paper we showed that the effects of radio interference between these technologies have been researched before, however not in a mobile communications system as comprehensive as our deployment. This has additional significance because in our system we can potentially control each of the sources of interference and attempt to minimise its negative effects through collaborative channel selection. Using our understanding of the interference problem domain we can now attempt to develop a channel selection scheme that attempts to identify radio spectrum in the 2.4GHz ISM band that is already saturated and then dynamically adapt its use. This concept in itself is potentially very challenging, bringing additional pitfalls because of the highly mobile nature of our deployment, but if solved correctly could provide an effective tool for other mobile networking solutions of this nature as and when they start to be deployed.

References

1. Morecambe Bay Search and Rescue, http://baysearchandrescue.org.uk/
2. McCarthy, B., Edwards, C., Dunmore, M.: Using NEMO to Support the Global Reachability of MANET Nodes. In: Proceedings of Twenty Eighth Annual Joint Conference of the IEEE Computer and Communications Societies (INFOCOM 2009), Rio de Janeiro, Brasil, April 19–24 (2009)
3. Montenegro, G., Kushalnagar, N., Hui, J., Culler, D.: Transmission of IPv6 Packets over IEEE 802.15.4 Networks. IETF RFC 4944 (September 2007)
4. Thubert, P., et al.: NEMO Basic Support Protocol. IETF RFC 3963 (January 2005)
5. McCarthy, B., Edwards, C., Varkliotis, S., Kirstein, P.: Incorporating mobile sensor networks into the internet - an all ip approach. In: Proceedings of the 4th International Workshop on Mobility in the Evolving Internet Architecture (MOBIARCH 2010). ACM, Chicago (2010)
6. Lippert embedded systems homepage, http://www.lippertembedded.de/index.php
7. The Contiki operating system, http://www.sics.se/contiki
8. Smartlife technology homepage, http://www.smartlifetech.com/
9. Nonin mdical inc, onyx ii 9560 wireless pulse oximeter homepage, http://nonin.com/PulseOximetry/Fingertip/Onyx9560
10. Handziski, V., Hauer, J., Wolisz, A.: Experimental study of the impact of wlan interference on ieee 802.15.4 body area networks. HotEmNets (2010)
11. Cypher, D., Golmie, N., Rebala, O.: Performance analysis of low rate wireless technologies for medical applications. Computer Communications 28(10), 1255–1275 (2005)
12. Howitt, I., Gutierrez, J.: Ieee 802.15.4 low rate - wireless personal area network coexistence issues. In: Wireless Communications and Networking, WCNC 2003, vol. 3, pp. 1481–1486. IEEE, Los Alamitos (2003)
13. Pollin, S., Hodge, B., Tan, I., Chun, C., Bahai, A.: Harmful coexistence between 802.15.4 and 802.11: A measurement-based study. In: 3rd International Conference on Cognitive Radio Oriented Wireless Networks and Communications, CrownCom 2008, pp. 1–6 (May 2008)

Visibility Levels: Managing the Tradeoff between Visibility and Resource Consumption*

Junyan Ma[1,2] and Kay Römer[2,3]

[1] School of Computer Science, Northwestern Polytechnical University, China
[2] Institute for Pervasive Computing, ETH Zurich, Switzerland
[3] Institute of Computer Engineering, University of Lübeck, Germany

Abstract. Pre-deployment tests of sensor networks in indoor testbeds can only deliver a very approximate view of the correctness and performance of a deployed sensor network and it is therefore common that after deployment problems and failures occur that could not be observed during pre-deployment tests. Finding and fixing such problems requires visibility of the system state, such that an engineer can identify causes of misbehavior. Unfortunately, exposing the internal state of sensor nodes requires resources such as communication bandwidth and energy: the better visibility of system state is required, the more resources are needed to extract that state from the sensor network. In this paper we propose a concept and tool that give the user explicit control over this tradeoff. Essentially, the user can specify a resource budget and our tool strives to provide best possible visibility while not exceeding the resource budget. We present the design of our vLevels framework and report the results of a case study demonstrating that the overhead of our approach is small and that visibility is automatically adjusted to meet the specified resource budget.

1 Introduction

Being deeply embedded into the real world, the function of sensor networks is heavily affected by their interaction with the environment. Therefore, pre-deployment tests in testbeds can only deliver a very approximate view of the correctness and performance of a deployed sensor network and it is therefore common that after deployment problems and failures occur that could not be observed during pre-deployment tests. Finding and fixing such problems is difficult due to limited access to the deployed network – both physically and due to constrained resources.

A key requirement for debugging deployed sensor networks is visibility of the system state, i.e., the ability of an engineer to observe the internal program state of the sensor nodes. Unfortunately, there is a fundamental tradeoff between visibility and resource consumption. As visibility requires communication to expose internal node states to an external observer, increasing visibility requires more resources. Since resources are

* This work has been partially supported by the Swiss National Science Foundation (NCCR-MICS, 5005-67322), the European Commission (CONET, FP7-2007-2-224053), and the National Key Technology R&D Program of China (2007BAD79B00).

P.J. Marron et al. (Eds.): REALWSN 2010, LNCS 6511, pp. 49–61, 2010.

scarce in sensor networks, a balance between a sufficient level of visibility and a tolerable consumption of resources needs to be found. This is especially true for situations where the system state needs to be observed over a long period of time in order to collect enough evidence to make an informed decision about the causes of observed problems.

While there exist a number of tools to give an observer visibility into internal node states, they do not offer a turning knob to the user to select a reasonable balance between visibility and resource consumption. With these existing tools, the user can define which states should be visible, and the resource consumption is implied by these requirements. However, it is very difficult for the user to predict the resulting resource consumption and hence almost impossible to exert fine-grained control over the permissible amount of resource consumption.

The goal of our work is to provide the user with an explicit turning knob to balance visibility and resource consumption. With our approach, the user can specify both visibility requirements and a resource budget and our system will provide best possible visibility while not exceeding the given resource budget. Here, visibility means the creation of traces of the system state that are either logged into memory for later post-mortem analysis, or transmitted over the wireless channel for online inspection. The resource budget is therefore defined in terms of available storage space or communication bandwidth. Our approach is embodied in a software framework called *vLevels*. The paper continues with the design of vLevels in Sect. 2, implementation aspects in Sect. 3, a case study in Sect. 4, and with a summary of related work in Sect. 5. Finally, we provide our conclusions in Sect. 6.

2 Design

vLevels offers visibility into the system state by creating snapshots of user-defined slices of the system state. The key innovation in vLevels is that it provides a turning knob for the user to specify a resource budget such that best possible visibility is offered while not exceeding this budget.

To realize this turning knob, vLevels offers three core abstractions: *visible objects* are slices of system state that should be made visible to an observer. When a user-define event occurs (such as changes of variable values or invocations of certain functions), a snapshot of the state of the visible object is logged. For each visible object, the user can define an ordered set of *visibility levels*. A visibility level is a user-defined lossy compression function that compresses a snapshot of the state of a visible object. Higher visibility levels result in more accurate snapshots which also consume more resources. Lower visibility levels result in less accurate snapshots which consume less resources. Finally, an *observation scheme* defines a resource budget and assigns priorities to visible objects, such that a scheduler can automatically select a visibility level for each visible object so as to maximize visibility while not exceeding the resource budget. We continue with an overview of the architecture of vLevels, before discussing the components of the architecture in detail.

2.1 System Architecture

We assume a traditional node-centric programming model, where developers write code in an imperative programming language such as C that is compiled, uploaded, and

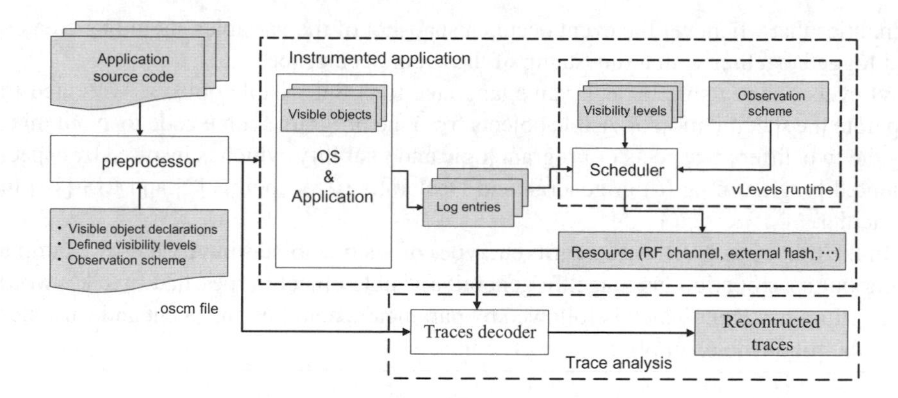

Fig. 1. System architecture of vLevels

executed at the sensor nodes. With vLevels, the user creates an additional *.oscm* input file that contains the specification of visible objects, visibility levels, and observation schemes. As illustrated in Fig. 1, both the source code and the .oscm file form the input to a preprocessor that modifies the source code such that snapshots of the state of visible objects are generated when certain events occur. The resulting source code is then compiled, linked with a vLevels runtime library, and uploaded to the sensor nodes.

When the application is executing on the sensor node, the instrumented code generates snapshots of the state of visible objects. These timestamped snapshots are then passed to a scheduler that dynamically selects an appropriate visibility level for each snapshot such that the resource budget specified in the observation scheme is not exceeded. The resulting compressed snapshots that constitute the traces of the system state are then either stored in (flash) memory or sent over the wireless channel. Using the compressed traces (that include timestamps and the chosen visibility levels) downloaded from the flash memory or received over the wireless channel, the original uncompressed traces of visible objects can be approximately reconstructed and analyzed by the user. The lower the visibility levels used for compression, the less accurate the reconstructed states will be. For the reconstruction, the compression schemes defined by the visibility levels are compiled by the preprocessor into appropriate decompression code that can be executed on the user's computer.

As the above discussion indicates, vLevels builds upon a number of fundamental services such as storage management (e.g., Contiki's Coffee filesystem [15]), timestamping and synchronization services as well as data collection (e.g., as described in [12]). Due to space constraint we focus on the innovative aspects of vLevels in this paper, rather than re-iterating those fundamental and well-understood services.

2.2 Visible Objects

The complete state of a program is typically too large to make it visible in its entirety. Therefore, a user has to specify which part of the program state is of interest to him. In vLevels, the *visible object* abstraction allows a user to specify the interesting state. Essentially, a visible object specifies an event and a set of program variables, with the semantics that whenever the event occurs a snapshot of the variables should be created and logged together with a timestamp of the event occurrence.

vLevels offers a simple declarative language to define visible objects. We opted to separate the specification of visible objects from the program source code to avoid mixing the two different aspects of program logic and visibility, which is inspired by aspect oriented programming [6] in general, and Declarative Tracepoints [2] and LIS [14] in particular.

In the following we illustrate different types of visible objects in vLevels by example using our specification language. In general, a visible object is specified by a keyword that defines the event which is followed by parameters detailing the event and variables as in the following examples:

```
1 var tracking.c::m_state
2 fvar tracking.c:sample_light_process:light
3 fhdr reports_mngmt.c:leader_report_update:reading x y
4 tpoint tracking.c:tp_leader_cycle:report_nr
5 var estate as example.c::m_state
```

The `var` keyword in line 1 defines a single global variable as a visible object, the event being any assignment of a different value to that variable. The parameters specify the name of the source file (`tracking.c`) and the name of the variable (`m_state`). The `fvar` keyword in line 2 is similar, but refers to a local variable (`light`) of a given function (`sample_light_process`) in the given source file.

Keywords `fhdr` (function header) in line 3 and `fftr` (function footer) also refer to local variables of a function including in particular the function parameters, but the triggering event is the invocation of the function (`fhdr`) or the return from the function (`fftr`). In the example in line 3 the parameters `reading`, `x`, and `y` of function `leader_report_update` are logged just before the first instruction of the function is executed.

The keyword `tpoint` (trace point) in line 4 generalizes this concept, where the triggering event is when control flow reaches a given point in the program. As the specification of line numbers is very error prone, we opted for inserting a marker comment into the source code, such that the snapshot of a given set of variables is generated whenever the control flow reaches this marker in the code. The code below shows such a marker comment, which has to follow the format `@visible tpoint` followed by the name of the tracepoint. This name (`tp_leader_cycle`) is also given in the specification of the visible object, followed by the names of the variables.

```
...
  target_pos_estimation(&target);
  /* @visible tpoint tp_leader_cycle */
  member_reports_clear();
...
```

For keywords `var` and `fvar`, variable names are used as their corresponding visible object names by default. Likewise, function names and tracepoint names are used as names of visible objects defined with `fhdr`, `fftr` and `tpoint`. The `as` keyword in line 5 is used to rename a visible object when there is a conflict.

2.3 Visibility Levels

Visibility levels are user-defined compression schemes to reduce the size of traces generated by visible objects. In particular, a set of visibility levels can be defined for each visible object. The levels in each set are ordered and numbered with small integers. Every level implements a specific tradeoff between accuracy of the snapshot of a visible object on the one hand, and size of the snapshot and therefore resource consumption on the other hand. Assuming there are N levels, then level N gives maximum visibility but also maximum resource consumption. By definition, level 0 produces empty output and thus gives zero visibility at zero cost. That is, both visibility and resource consumption monotonically increase with increasing level numbers. We will explain later in the paper how the visibility levels of a set of visible objects can be automatically selected such that a given resource budget is not exceeded.

Each visibility level essentially constitutes a lossy compression function that is defined by the user in a declarative manner using a number of basic operators. Moreover, the visibility levels of a visible object form a pipeline: The output of level k forms the input of level $k-1$. Level N is typically the raw snapshot of a visible object. The rationale for this design will become clear in Sect. 2.5, where we describe the scheduler that selects the visibility levels. The basic operation of the scheduler is to incrementally decrease the visibility levels of a snapshot by one. The pipelined design of visibility levels makes this a very efficient operation as the raw snapshot does not have to be saved in order to change the visibility level later.

The left part in Table 1 shows the definition of a set of three visibility levels for a program variable `light` that has been declared a visible object as described in Sect. 2.2 (line 2 in the example there). The example declares three visibility levels, the declarations of which are separated by blank lines. Each visibility level consists of zero or more compression operations (one per line) followed by a single output operation `log` to produce the output.

Visibility level 3 is declared first in line 3, followed by level 2 in lines 5 and 6, and finally level 1 in line 8. Level 3 just outputs the raw snapshot of variable `light` using the `log` operator. Level 2 uses the `filter` compression operator to ignore all snapshots where the `light` value is ≤ 200. The remaining `light` values are output with `log`. Level three takes the output of level 2 as input, but instead of outputting the light value, it just creates an empty log entry using `log` without parameters whenever a new value > 200 is assigned to variable `light`. As log entries do also contain a timestamp, this is an indicator of *when* the variable has been assigned new values > 200, but not which values. As we can observe in this example, the visibility of the `light` variable as well as the size of the generated log entries are monotonically increasing with increasing level number.

Besides `log` and `filter`, vLevels also provides `remapper` and `bits`. Operator `bits` (bits selector) selects a subset of the bits of its input using a bit mask. `bits` is often used to deal with network messages to extract the relevant protocol fields. Operator `remapper` remaps a set of integer values to a new set of integer values. Essentially, this operator reduces the precision of a scalar value to a smaller number of bits.

We conclude this section with a discussion of the design rationale behind visibility levels. Essentially, with vLevels a user can define custom lossy compression schemes, allowing him to exploit his domain knowledge about what is important state (and should not be lost during compression) and what is not. This is especially important for the compression of systems logs containing very different data types (e.g., outputs of

Table 1. Visibility levels and observation scheme

Visibility Levels	Observation Scheme
	1 oscheme leader_algorithm
	2 {
1 vlevels lreading for light	3 budget = 60 BPS
2 {	4
3 log light	5 light@levels = lreading
4	6 light@priority = 2
5 filter light > 200	7 light@rpolicy = DEE
6 log light	8
7	9 election_timeout@levels = electmout
8 log	10 election_timeout@priority = 1
9 }	11 election_timeout@rpolicy = DEE
	12 }

different sensor types, program state variables). An alternative design could use general purpose compression algorithms. However, they are often optimized for specific data types and do not allow to use to incorporate domain knowledge during compression.

2.4 Observation Schemes

An observation scheme specifies how to select the visibility levels of a set of visible objects such that a certain resource budget is not exceeded. In addition to the resource budget itself, the observation scheme also defines policies how to prioritize visible objects among each other, and how to prioritize different snapshots of a single visible object among each other. In the case that the resource budget is not sufficient to log all snapshots of all visible objects at maximum visibility level, these policies control the selection of visible objects whose visibility level needs to be lowered. Let us consider the observation scheme named `leader_algorithm` in the right part of Table 1.

The `budget` keyword is used to define a bandwidth budget of 60 bytes per second (BPS), meaning that the log data is transmitted online over a communication channel and the log data bandwidth should not exceed 60 BPS. Based on the application requirements, the user decides how much resources can be spent for monitoring and selects the budget appropriately. The budget can also be changed later during runtime. As the radio is the dominating energy sink, with a simple calibration step it is possible to map bandwidth to approximate energy consumption, such that instead of specifying a bandwidth budget, one can also specify an energy budget. The system also supports storage of log data in flash memory for later offline analysis. In the latter case, a storage budget is specified in units of bytes (B), kilobytes (KB), or megabytes (MB).

The example observation scheme considers two visible objects: `light` and `election_timeout`. By assigning the name of a visibility level set to the `@levels` attribute, a visibility levels set can be selected for a visible object as in lines 5 and 9. The `@priority` attribute (lines 6 and 10) defines the relative importance of the visible objects, where smaller values mean more importance. If there are not enough resources to log both objects at maximum visibility level, then preference (i.e., higher level) will be given to the object with the smallest priority (here, the `election_timeout` object).

However, the visibility level can also change dynamically among different snapshots of the same visible object. The `@rpolicy` attribute (lines 7 and 11) defines the policy how to prioritize different snapshots of the same visible object among each other. For example, the DEE (Drop Earliest Entry) default policy specifies that priority should be given to the latest snapshots, i.e., the earliest snapshot should be dropped first if the budget is not sufficient. Other supported policies are DLE (Drop Latest Entry) and MMTIU (Minimize Maximum Time Interval between Updates). The latter policy drops the snapshot that results in the minimum increase of time gap between successive snapshots in the trace, which is suitable for signal reconstruction if the visible object represents the output of a sensor.

2.5 Scheduler

The scheduler is the component in our system that dynamically selects visibility levels for visible objects such that the given bandwidth/energy or storage budget is not exceeded. It should be noted that in our system the budget is considered a user-defined constant and the scheduler assumes this budget is always available during runtime. The key idea is that the scheduler maintains a buffer of a fixed size and enters new log entries at the end of the buffer. If the remaining space in the buffer is not sufficient to hold the new entry, then one or more existing log entries in the buffer are selected for reduction of their visibility level.

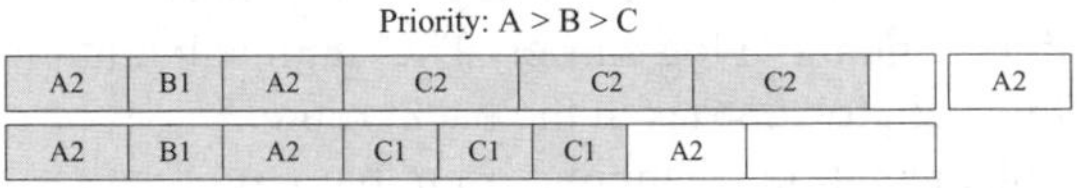

Priority: A > B > C

A2	B1	A2	C2	C2	C2		A2
A2	B1	A2	C1	C1	C1	A2	

Fig. 2. Examples of the scheduler algorithm

In case of logging into (flash) memory for later offline analysis, this buffer equals the storage space in the memory. An efficient file system such as Coffee [15] is required to manage access to the flash. In case of online monitoring, the contents of the buffer are sent over the communication channel at regular intervals, such that this interval equals the buffer size divided by the bandwidth budget. As transmission of the buffer contents is not instantaneous, a second buffer is used. While one of the buffers is being filled with log entries, the contents of the other buffer are transmitted in the background.

Both in the online and offline modes the basic operation of the scheduler is as follows. When appending a new log entry to the buffer, it is initially inserted with the current visibility level of the visible object. If the available space in the buffer is not sufficient to hold the new entry, then the visibility level of the object with the lowest priority is reduced by one until there is enough space in the buffer to hold the new entry. The core operation of the scheduler is hence to reduce the visibility level of a log entry in the buffer by one until it eventually reaches zero, i.e., the log entry disappears. This is also the reason for the pipelined design of the visibility levels as described in Sect. 2.3, since reducing the visibility level of an entry in the buffer from k to $k-1$ is then efficiently implemented by applying the compression function of level $k-1$ to the entry. If the lowest priority object is the object associated with the new log entry and its level is already 1, then a log entry is selected to be replaced with the new one according to the replacement policy. Figure 2 gives an example of how the scheduler works. The notation A2 stands for a log entry of visible object A with visibility level 2. As shown in the figure, a new entry A is added but there is not enough space in the buffer. As C is the visible object with the lowest priority, the visibility level of all C entries is reduced from 2 to 1.

3 Implementation

Our prototype implementation of vLevels on the Contiki operating system consists of two main parts: a preprocessor and a runtime system.

The *preprocessor* reads the C source files and the `.oscm` file containing the specification of visible objects, visibility levels, and observation schemes. In particular, the preprocessor assigns a unique identifier to every visible object to identify log entries; source code is instrumented to log snapshots of visible objects; compression functions are generated for visibility levels; parameters from the observation schemes are extracted and passed to the scheduler; and a trace decoder is also generated for reconstructing traces from the collected logs. The instrumentation part is implemented using CIL (C Intermediate Language) [9]. Operating on the intermediate representation of C programs generated by CIL, code analysis and transformation are performed, such as code injection after an assignment to a visible variable.

The *runtime* mainly consists of the scheduler and a module for storage or wireless transmission of buffers containing log entries. The runtime maintains a separate thread for sending off buffers with log entries in the background.

In our current implementation, changing .oscm file requires to preprocess and compile the code and to upload the new image to the sensor nodes. However, through binary instrumentation techniques [2] and run-time dynamic linking [3], it would also be possible to insert new visible objects, update modules and load new modules into an executing program without losing its state. We leave this aspect for future work.

4 Case Study

To verify the feasibility of our design, we conduct a preliminary experiment on applying vLevels to a target tracking application similar to EnviroTrack [1]. We investigate memory overhead, runtime overhead, as well as the impact of buffer size on observation accuracy and bandwidth throttling. We choose Tmote sky as our sensor node hardware platform and Contiki [4] for the operating system running on the nodes. The experiment is carried out using the cycle-accurate COOJA/MSPSim simulator [5].

4.1 Tracking Application

We consider a simplified single-target tracking application where sensor nodes can sense the proximity of the target (e.g., using an IR light sensor to detect the presence of living beings). If the sensor reading is above threshold the target is considered detected. One of the nodes close to the target is elected as the leader and all detecting nodes send messages with their sensor values and locations to the leader which computes the target location and notifies a sink. When the target moves away from the the leader, the leader role is handed over to a closer node. The tracker is implemented as a state machine with states *idle* (no target detected), *leader*, *candidate* (target detected but not a neighbor of the leader), *member* (target detected and neighbor of a leader), *sentry* (no target detected but neighbor of leader), or *temporary* (a leader that lost the target). A candidate turns into leader if it did not receive an announcement from another leader during a certain timeout.

4.2 Visibility Specification

In our experiment, four visible objects are declared to observe the execution of the application. The first visible object is the state `m_state` of the tracker with three visibility levels: level 3 (the original value of every assignment to the variable), level 2 (one bit indicating if the state is unstable (i.e., *candidate* or *temporary*) or stable (i.e., remaining states)), level 1 (empty log entries indicating the assignment of the state variable). The second visible object is the local variable `light` in function `sample_light_process`, holding the most recent sensor reading. There are three visibility levels: level 3 (original variable values), level 2 (values greater than the detection threshold), level 1 (empty log entries indicating assignment of values greater than the threshold). The third visible object is the invocation of `election_timeout`, the timeout callback function for leader election. There is only one visibility level (empty log entries indicating the invocation of the function). The last visible object is `tp_leader_cycle`, a tracepoint in the target estimation function of the leader creating snapshots of four variables: the number of report messages received from members, the average sensor reading computed by leader, and the estimated (x, y) location of the target. There are four visibility levels: level 4 (values of all variables), level 3 (number of reports, the most significant 4 bits of the target position (x, y), the most significant 8 bits of the average sensor reading), level 2 (number of reports), level 1 (empty log entry indicating the hit of the tracepoint).

The observation scheme specifies a bandwidth budget of 4 bytes/s. The priorities of visible objects `m_state` and `election_timeout` are set to 1, the priority of `tp_leader_cycle` is set to 2, and the priority of `light` is set to 3. DDE is selected as replacement policy for all objects.

4.3 Memory Overhead

We investigate the memory (RAM, ROM) overhead of vLevels by compiling the output of the preprocessor with msp430-gcc version 3.2.3 with the `-Os` optimization switch using a June 18, 2010 CVS snapshot of Contiki.

Table 2 shows memory footprint a) without vLevels, b) with vLevels and 0 byte buffers and no visible objects, and c) with vLevels and 50 byte buffers and all four visible objects, respectively. b) results in an increase of RAM by 70 bytes and an increase of ROM by 2608 bytes. As vLevels maintains two buffers, the two 50 byte buffers in c) result in an increase of RAM by 100 bytes. Besides buffer overhead, an additional 10 bytes of RAM are required for each visible object. Additional visibility levels result in extra ROM consumption. Also, instrumentation for snapshot creation results in ROM overhead. One logging instrumentation for a 16-bit variable consumes about 40 bytes. The tracker state variable is assigned at 15 places, resulting in a ROM overhead of 600

Table 2. vLevels memory footprint for Contiki on Tmote sky

Binary	RAM (bytes)	ROM (bytes)
Tracking application	5956	29510
+vLevels 0B	6026	32118
+vLevels 50B	6166	33808

bytes. In summary, the overhead of vLevels is about 2% of the total RAM and 8% of the total ROM of a Tmote Sky for the studied application, which we find to be acceptable.

4.4 Runtime Overhead

To evaluate the runtime behavior of vLevels, we run the application with vLevels in the COOJA/MSPSim simulator (CPU speed 3.9 MHz). We only consider a single tracking node. The experiment lasts 300 seconds, the target appears at 50 seconds, remains static, and disappears at 250 seconds. We measure the vLevels overhead in terms of CPU cycles for initialization (call to `vlevel_init`), logging (invocation of `vlevel_log`), the buffer management thread (`buf_proc`), and the send thread excluding the actual transmission of the data (`logcast_proc`).

Table 3. Average cycle counts of different parts of vLevels

buffer size	vlevel_init	vlevel_log	buf_proc	logcast_proc	total	ratio
12	1716	1056	803	1803	489378	0.04%
24	1716	821	799	1775	413061	0.04%
48	1716	866	791	1723	390942	0.03%

Table 3 shows the average cycle counts of these parts of vLevels during the experiment for different buffer sizes. `vlevel_log` is invoked 130 times in total and its cycle count varies with buffer size. Small buffers result in higher overhead as the scheduler must downgrade visibility levels more frequently to fit everything into the buffer. The overheads of `buf_proc` and `logcast_proc` are independent of the buffer size, but `buf_proc` is called more often for smaller buffers. The number of invocations to `logcast_proc` is proportional to the bandwidth constraint given by users. Finally, we calculated the aggregate overhead introduced by vLevels during the complete experiment and the ratio of the aggregate overhead to the total CPU cycles as shown in Table 3, which we find to be acceptable.

4.5 Accuracy and Bandwidth

Observation is the more accurate, the higher the selected visibility levels. Figure 3 depicts the visibility levels selected by the scheduler over time for the four visible objects for different buffer sizes. The figure shows that the visibility level of `light` degrades as log entries for objects with higher priority (i.e. `m_state`, `election_timeout` are generated between time 50 and 250 seconds. Note that higher-priority objects are assigned higher visibility levels in case of larger buffer sizes as there is a bigger set of log entries to pick from when reducing visibility levels. Hence, larger buffers result in more

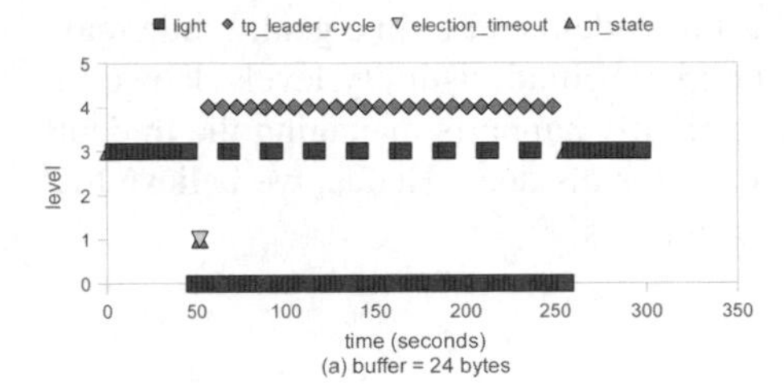

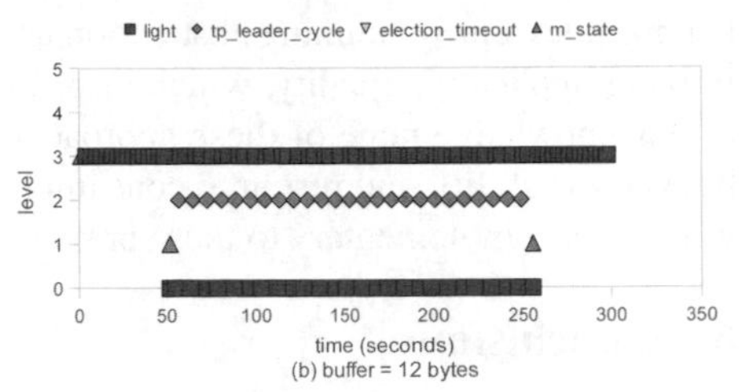

Fig. 3. Accuracy of collected data for different buffer sizes

Table 4. Throttled bandwidth and reporting latency by buffer sizes

	original data	48B buffer	24B buffer	12B buffer
bandwidth [bytes/sec]	5.53	3.5	3.24	3.94
latency [sec]	-	12	6	3

accurate observations at the cost of a longer reporting latency for the same bandwidth budget.

Table 4 shows the bandwidth of the uncompressed raw logging data and the bandwidth throttled by vLevels for different buffer sizes during the experiment. The numbers show that the throttled bandwidth is always lower than the bandwidth budget specified in the observation scheme (i.e., 4 bytes/s).

5 Related Work

In the recent past, several techniques and tools for monitoring and debugging deployed sensor networks have been proposed. Initial efforts towards debugging sensor networks are Sympathy [10] and Memento [13], both of which support the detection of a fixed set of problems. To improve repeatability of distributed event-driven applications, EnviroLog [8] provides an event recording and replay service that helps users to debug problems that are difficult to reproduce. Tools such as Marionette [17] support logging and retrieval of runtime state for fault diagnosis. Clairvoyant [18] is an interactive source-level debugger for sensor nodes. Declarative Tracepoints [2] implements a language abstraction for insertion of tracepoints into binary program images that bears some similarity to our notion of visible objects. To minimize the interference with debugged sensor networks, tools such as SNIF [11] that diagnose networks by analyzing sniffed messages can detect many failure symptoms, but visibility is limited to network messages. By allowing nodes to broadcast program state, Passive Distributed Assertions [12] support the detection of bugs caused by the incorrect interaction of multiple nodes. By optimizing log collection, LIS [14] enables users to extract detailed execution traces from resource-constrained sensor nodes. In [16], visibility is regarded as an important metric for protocol design and is improved by creating an optimal decision tree so that the energy cost of diagnosing the cause of a failure or behavior can be minimized. Unlike providing a principle for protocol design, our approach creates a mechanism to tune the visibility of internal node states. Outside the context of debugging, Energy Levels

[7] provides a programming abstraction to meet user-defined lifetime goals while maximizing application quality, which inspires the idea behind visibility levels. However, to our knowledge none of these approaches explicitly supports managing the tradeoff between visibility and resource consumption as vLevels does. Hence, we believe that vLevels is complementary to these previous techniques.

6 Conclusions

Debugging deployed sensor networks requires visibility of the node states. However, increasing visibility also incurs a higher resource consumption in terms of communication bandwidth or storage space. Especially for long-term monitoring of a sensor network it is hence crucial to find the right balance between sufficient visibility and tolerable resource consumption. Existing monitoring tools lack the ability to explicitly manage this tradeoff. We address this limitation by proposing vLevels, a framework that allows the user to specify a resource budget and the runtime provides best possible visibility into the system state while not exceeding the resource budget. By means of a case study of a tracking application we showed that the memory and runtime overhead of vLevels is reasonably small and that vLevels can automatically adjust visibility to meet the resource budget.

References

1. Abdelzaher, T., Blum, B., Cao, Q., Chen, Y., Evans, D., George, J., George, S., Gu, L., He, T., Krishnamurthy, S., Lou, L., Son, S., Stankovic, J., Stoleru, R., Wood, A.: Envirotrack: Towards an environmental computing paradigm for distributed sensor networks. In: Proc. ICDCS 2004, pp. 582–589. IEEE Computer Society, Washington (2004)
2. Cao, Q., Abdelzaher, T., Stankovic, J., Whitehouse, K., Luo, L.: Declarative tracepoints: a programmable and application independent debugging system for wireless sensor networks. In: Proc. SenSys 2008, pp. 85–98. ACM, New York (2008)
3. Dunkels, A., Finne, N., Eriksson, J., Voigt, T.: Run-time dynamic linking for reprogramming wireless sensor networks. In: Proc. SenSys 2006, pp. 15–28. ACM, New York (2006)
4. Dunkels, A., Gronvall, B., Voigt, T.: Contiki - a lightweight and flexible operating system for tiny networked sensors. In: Proc. LCN 2004, pp. 455–462. IEEE Computer Society, Washington (2004)
5. Eriksson, J., Österlind, F., Finne, N., Tsiftes, N., Dunkels, A., Voigt, T., Sauter, R., Marrón, P.J.: Cooja/mspsim: interoperability testing for wireless sensor networks. In: Proc. SIMUTools 2009, pp. 27–27. ICST, Brussels (2009)
6. Kicales, G., Lamping, J., Mendhekar, A., Maeda, C., Lopes, C., Loingtier, J.M., Irwin, J.: Aspect-oriented programming. In: Aksit, M., Matsuoka, S. (eds.) ECOOP 1997. LNCS, vol. 1241, Springer, Heidelberg (1997)
7. Lachenmann, A., Marrón, P.J., Minder, D., Rothermel, K.: Meeting lifetime goals with energy levels. In: Proc. SenSys 2007, pp. 131–144. ACM, New York (2007)
8. Luo, L., He, T., Zhou, G., Gu, L., Abdelzaher, T.F., Stankovic, J.A.: Achieving repeatability of asynchronous events in wireless sensor networks with envirolog. In: Proc. INFOCOM 2006, pp. 1–14. IEEE Press, New York (2006)
9. Necula, G.C., Necula, G.C., McPeak, S., Rahul, S.P., Weimer, W.: CIL: Intermediate language and tools for analysis and transformation of c programs. In: Parra, G. (ed.) CC 2002. LNCS, vol. 2304, pp. 209–265. Springer, Heidelberg (2002)

10. Ramanathan, N., Chang, K., Kapur, R., Girod, L., Kohler, E., Estrin, D.: Sympathy for the sensor network debugger. In: Proc. SenSys 2005, pp. 255–267. ACM, New York (2005)
11. Ringwald, M., Römer, K., Vitaletti, A.: Passive inspection of sensor networks. In: Aspnes, J., Scheideler, C., Arora, A., Madden, S. (eds.) DCOSS 2007. LNCS, vol. 4549, pp. 205–222. Springer, Heidelberg (2007)
12. Römer, K., Ma, J.: PDA: Passive distributed assertions for sensor networks. In: Proc. IPSN 2009, pp. 337–348. IEEE Computer Society, Washington (2009)
13. Rost, S., Balakrishnan, H.: Memento: A health monitoring system for wireless sensor networks. In: IEEE SECON 2006, pp. 575–584. IEEE Press, New York (2006)
14. Shea, R., Srivastava, M., Cho, Y.: Lis is more: Improved diagnostic logging in sensor networks with log instrumentation specifications. Tech. Rep. TR-UCLA-NESL-200906-01 (June 2009)
15. Tsiftes, N., Dunkels, A., He, Z., Voigt, T.: Enabling large-scale storage in sensor networks with the coffee file system. In: Proc. IPSN 2009, pp. 349–360. IEEE Computer Society, Washington (2009)
16. Wachs, M., Choi, J.I., Lee, J.W., Srinivasan, K., Chen, Z., Jain, M., Levis, P.: Visibility: a new metric for protocol design. In: Proc. SenSys 2007, pp. 73–86. ACM, New York (2007)
17. Whitehouse, K., Tolle, G., Taneja, J., Sharp, C., Kim, S., Jeong, J., Hui, J., Dutta, P., Culler, D.: Marionette: using rpc for interactive development and debugging of wireless embedded networks. In: Proc. IPSN 2006, pp. 416–423. ACM, New York (2006)
18. Yang, J., Soffa, M.L., Selavo, L., Whitehouse, K.: Clairvoyant: a comprehensive source-level debugger for wireless sensor networks. In: Proc. SenSys 2007, pp. 189–203. ACM, New York (2007)

Flexible Online Energy Accounting in TinyOS

Simon Kellner

System Architecture Group
Karlsruhe Institute of Technology
kellner@kit.edu

Abstract. Energy is the most limiting resource in sensor networks. This is particularly true for dynamic sensor networks in which the sensor-net application changes its hardware utilization over time. In such networks, offline estimation of energy consumption can not take into account all changes to the application's hardware utilization profile and thus invariably returns inaccurate estimates. Online accounting methods offer more precise energy consumption estimates. In this paper we describe an online energy accounting system for TinyOS consisting of two components: An energy-estimation system to collect information about energy consumption of a node and an energy-container system that allows an application to collect energy-consumption information about its tasks individually. The evaluation with TinyDB shows that it is both accurate and efficient.

Keywords: energy accounting policy tinyos.

1 Introduction

Energy still is the most critical resource in sensor networks. Limitations on energy supply as well as on other resources have led to operating system designs that offer only minimalistic hardware abstractions. The core of TinyOS, for example, is an event-based system that helps application developers in dealing with asynchronous hardware requests, and little else. One effect of this design decision is to make developers more considerate about hardware usage and therefore energy consumption. TinyOS makes it hard to actively wait for a hardware event to occur, while making it easy to react to the same event, which is the more energy-efficient approach in most situations.

One approach to designing sensor-net applications that meet pre-defined energy consumption requirements is to develop an application whose hardware utilization pattern is simple enough to allow predictions on the application's energy consumption. Global parameters of such applications can then be changed to accommodate energy consumption requirements. But the lack of convenient hardware abstractions does not necessarily limit developers in creating complex applications. A sensor network running the TinyOS-based TinyDB application, for example, allows users to issue (SQL-like) queries to the sensor network at a time of their choosing. Planning the energy consumption of nodes in this network can not be done a-priori, because the energy consumption characteristics of a node running TinyDB change with the queries it processes.

P.J. Marron et al. (Eds.): REALWSN 2010, LNCS 6511, pp. 62–73, 2010.

Control of energy consumption in this scenario is only feasible using online energy accounting on the sensor nodes. Information on the energy consumption of whole nodes, however, does not offer much information. An energy-intensive query might, for example, be only revealed by comparing node energy consumption before and after a query was sent into the network. Energy consumption of queries, on the other hand, can be readily used to decide if a query consumes too much energy and has to be canceled before it wears down the energy supplies of the sensor network.

This paper makes the following contributions:

- An online energy-estimation system for TinyOS that allows sensor nodes to become aware of their energy consumption.
- An energy container system for TinyOS that allows application developers to collect energy-consumption information about control flows in the application.
- A set of accounting policies that can be used to adapt the energy-container system to its purpose as set by the application developer.

The paper is structured as follows: After presenting related work in Sect. 2 we define a usage scenario in Sect. 3 that will be referenced later on. Then we present the design and selected implementation issues of the energy estimation system (Section 4) and the energy container system (Section 5). Section 6 details several accounting policies of our energy-container system. Following an evaluation of our systems in Sect. 7 we conclude with an outline of future work in Sect. 8 and closing remarks in Sect. 9.

2 Related Work

Management of energy in sensor networks has received significant attention in research over the last years, as it concerns the primary resource of such networks.

PowerTOSSIM [7] is similar to our own energy estimation system. It instruments OS components or simulations thereof to track power states and uses an energy model to compute energy consumption for one or more sensor nodes. PowerTOSSIM, however, targets off-line simulation, whereas our instrumentation and model are designed to be used in on-line energy accounting.

AEON [5] is the energy model used in the Avrora [8] simulator. It models the hardware's power states of a MICA2 node. Our energy model is based primarily on the MICAz node and additionally considers transitions between hardware states.

Schmidt, Krämer et al. [6] present another energy model used to make existing simulators energy-aware. Although they mention the potential to use their energy model in online energy estimation, they do not elaborate on that option further.

Dunkels et al. [2] present an energy-estimation system for the Contiki OS. This system is used to estimate energy consumption per hardware components. We employ a similar energy-estimation system and extend it with energy containers to a full energy-accounting system that is able to account energy based on control flows, which may span multiple hardware components.

Quanto [3] is an energy profiling mechanism for TinyOS that accounts energy consumption information of activities in an application. It employs a hardware energy meter to measure the total energy consumption of a sensor node, and tries to break this information down to energy consumption of individual hardware components. Our energy estimation system does not require any hardware instrumentation. It also provides more options for accounting policies, facilitating more use cases than energy profiling.

Resource Containers [1] are an abstraction in an operating system (OS) introduced by Banga, Druschel and Mogul providing flexible, fine-grained resource accounting on web-servers. The main idea is to separate execution (processes) from resource accounting, so that an application itself can define the entity being subject to accounting. In operating systems featuring CPU abstractions such as threads or processes, Resource Containers give administrators and users the ability to account all activity connected to a user request, which usually has a higher significance than process-based accounting. We adapt this concept to TinyOS and focus solely on energy as a resource. Consequently, we call our containers *energy containers*.

3 Scenario

In this paper we use the following reference scenario. A network of sensors is programmed with a dynamic application like TinyDB. We assume multiple users of the sensor network who periodically retrieve data from the network. They retrieve data by injecting queries into the network, which are then periodically processed by the application on the sensor nodes until a user stops them. The sensor-net application can process multiple queries concurrently over a long period of time.

Network operators and users should be able to intervene in the query processing to save energy.

4 Online Energy Estimation

An important part of the energy accounting system is the on-line estimation of a sensor node's energy consumption.

We recognize a sensor node as a collection of simple, independent hardware components controlled by one microcontroller (MCU). Therefore, we model a node's energy consumption using a collection of small state machines, one for each independent hardware component.

These state machines have a state for each distinguishable hardware power state, i.e., a hardware state exhibiting a characteristic current draw. Each state s is annotated with this current draw I_s. Transitions t in this state machine are annotated with the amount of electric charge Q_t they consume.

To compute an estimation of the energy a hardware component has consumed, we record the time T_s the hardware spent in each state s as well as the number of times N_t each transition t occurred and compute the estimated consumed energy E as

$$\frac{E}{V} = \sum_s T_s I_s + \sum_t Q_t N_t \ . \tag{1}$$

This is the same idea Dunkels et al. [2] use in their online energy estimation for the Contiki operating system.

We implemented the online energy estimation method presented above in TinyOS 2 for the MICAz platform. At the time of writing, we have energy estimation implementations for the ATmega128 microcontroller, the CC2420 radio chip, the LEDs, and the magnetometer on the MTS300 sensor board.

5 Energy Container System

To store the estimated energy usage per query, we employ a hierarchy of energy containers. With multiple energy containers in the system (e.g., for concurrent queries), we need some help from the application (e.g., TinyDB) to map activities to energy containers. Our energy-container system keeps this association intact.

In this section we will first present energy-container types and their structure in our system. We then will describe the way in which the application should interact with the energy container system. Afterwards we will detail how our system keeps energy-container associations intact.

5.1 Energy Container Structure

In our system, energy containers are hierarchically structured, as shown in Fig. 1.

Most containers in our energy container system are under the control of the application developer. They can be allocated, read, and released at the application's discretion. The application can also switch between containers, indicating that subsequent computations should be accounted to a different container.

In addition to these *normal containers*, the *root container* holds the energy consumption of a whole sensor node: All energy consumption is accounted both to the container selected by the application and to the root container. Together, these two types of containers form a flat hierarchy.

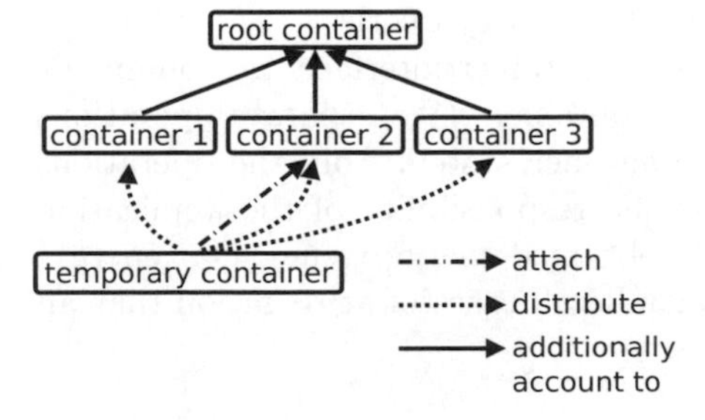

Fig. 1. Hierarchy of energy containers

```
ec_id newContainer();
void attachToContainer(ec_id id);
void switchToContainer(ec_id id);
uint32_t getContainer(ec_id id);
uint32_t getRootContainer();
void stopMonitoring(ec_id id);
```

Fig. 2. Interface to the container system

As a special case, a *temporary container* is used in cases where the application cannot know (yet) to which container the current energy consumption should be accounted. On a sensor node such a situation only occurs whenever a message is received by the radio: The message can belong to the query currently active, lead to the creation of a new query, or it could not be associated to query processing at all. For these cases we use one extra container that is activated upon reception of a message and is treated in a special way: When the message is found to belong to a known query, or creates a new one, the application has to *attach* the temporary container to the normal container used for that query. In this case the contents of the temporary container are added to the normal container, the currently active container is set to the normal container, and the temporary container is deactivated. If the application does not attach the temporary container to a normal one, our container system apportions the energy accounted to the temporary container among all currently active containers at the end of the message reception handler routine.

5.2 Energy Container Interface

Figure 2 shows the interface an application should use to work with our energy container system. The commands are ordered as they would be used in an application.

Upon reception of a query message, the application attaches itself to a known container (`attachToContainer`) or to a newly created one (`newContainer`). Before an application starts a processing step of a query, it switches to the container created for that query (`switchToContainer`). When creating a query response message, the application may choose to include the contents of one or more energy containers (`get{,Root}Container`). If a query should no longer be processed (i.e., removed from the system), it invokes `stopMonitoring` to completely deactivate the indicated container.

The presented interface intentionally does not provide full control over energy containers. For instance, there is no command to deactivate the currently active container without either removing it completely from the system or switching to another container. In our opinion, every action on a sensor node should be accountable to a request made by a user of a sensor network. Nevertheless it is possible to create new containers, for example, to account a maintenance task that is run on a sensor node and is independent of any queries.

Also intentionally absent from the energy container interface are commands to operate on the contents of the energy containers. We separate operations on energy containers within our energy container system from the operations on energy values in the application. It is the responsibility of the application to make network-wide use of these locally obtained energy values. Our energy-container system can not automatically handle all cases of aggregation that an application may perform on energy values.

5.3 Energy Container Implementation

As the previous section on the energy container interface shows, the application has to be modified to use energy containers. Naturally we strive to require as few changes to the application as possible, which means that our system has to keep associations to energy containers intact.

Resource containers in UNIX are attached to threads, so thread control blocks would be used to store references to associated resource containers, and the reference to the currently active thread would be used to access the currently active resource container. TinyOS, however, does not provide a CPU abstraction such as threads. TOSthreads, a TinyOS library providing threads, is optional and many applications do not require it, including TinyDB. TinyOS at its core therefore lacks a structure like a thread control block and a reference to the currently active thread.

In the absence of threads, we have to implement container associations in a different way: We use a set of TinyOS components to track the control flow of an application and to keep an energy container associated to this control flow. By *control flow* we mean a series of actions (instruction execution, hardware operation) where every action is a direct consequence of the actions preceding it. For example, a control flow to sample a value from a sensor can comprise: issuing a `read()` call, turning on the sensor, configuring it, reading it, turning the sensor off, and returning the value to the application in a `readDone` event.

A control flow can be suspended several times during its course through TinyOS, but all these suspensions stem from two cases: software queues and hardware operations. We say that a control flow is suspended when a piece of code performs an enqueue operation, and that it is resumed on the related dequeue operation. Similarly, we say that a control flow is suspended when it starts a hardware operation, and that it is resumed when the hardware causes an interrupt handler to be executed.

Software queues are frequently used in TinyOS. For example, instances of software queues are the queue of timer events in virtual timers, message output queues of communication modules, access request queues of shared resources, and the scheduler queue.

We instrumented several TinyOS components to send information about enqueue-, dequeue-, and hardware operations to our subsystem for control flow tracking. For software queues we implemented shadow queues of energy containers. During the enqueue operation, the shadow queue enqueues a reference to the currently active energy container, thereby associating the object being enqueued with this energy container. When an object is dequeued, the corresponding energy container from the shadow queue is activated, resuming the control flow with the energy container association intact. Control flow tracking over software queues is illustrated in Fig. 3.

Figure 4 shows that control flow tracking in hardware is handled differently. Not only is there just one energy container association that has to be stored, but, more importantly, the concurrency of hardware operations and program execution on the microcontroller means that there can be multiple active energy

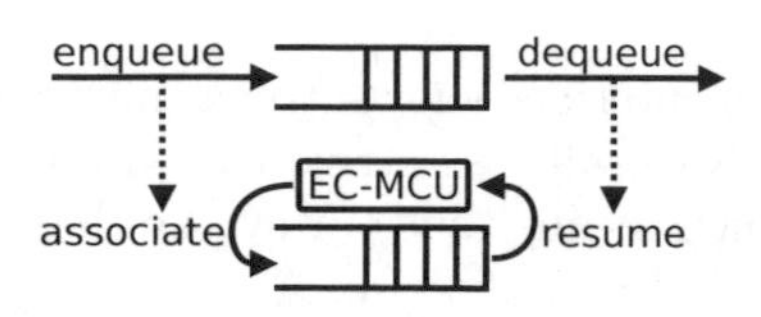

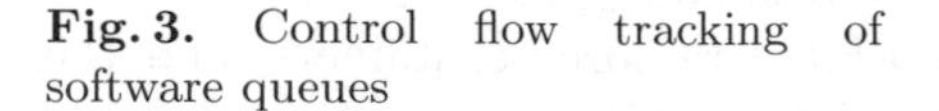
Fig. 3. Control flow tracking of software queues

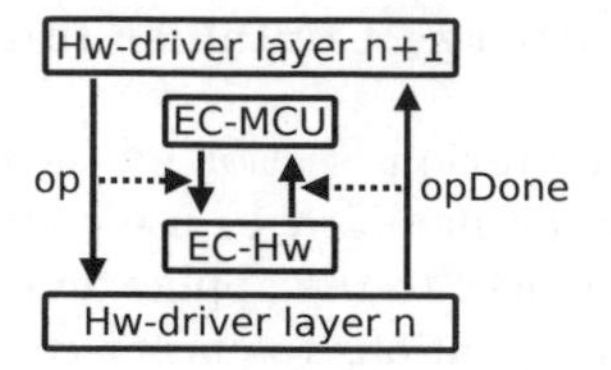

Fig. 4. Control flow tracking of hardware operations

containers on one node, each one associated with a different hardware component. Thus, our control-flow subsystem copies the current energy container association from the microcontroller to the hardware, and copies the association back when a hardware signal is received.

Altogether, the control flow tracking components ensure that an application has to switch containers only inside its own components, and only on switching query processing from one query to another.

6 Accounting Policy

Energy consumed during use of a hardware component is accounted to the energy container associated with the active control flow on the hardware. But energy is also consumed by the hardware before and after use: On startup, on shutdown, and between uses. We call this kind of energy consumption *collateral.*

There are many ways in which collaterally consumed energy can be accounted. The choice between these ways depends on the hardware usage pattern of the operating system, as the usage pattern defines whether collaterally consumed energy can be shared or not. It also depends on the reason why energy accounting is used: For energy profiling purposes, for example, an application developer might prefer not to account collaterally consumed energy at all, or account it to a separate container. A provider of a TinyDB network, however, might prefer to have all energy consumption accounted to TinyDB queries in a fair manner. As energy profiling systems already exist, we focus on a fair energy accounting in the TinyDB scenario. We identify hardware usage patterns and choose a suitable apportioning policy.

6.1 Single Use

The hardware usage pattern of the microcontroller (MCU) is simple: Its startup overhead is negligible, and then there is only one active control flow at a time. The apportioning policy used on energy consumed by the MCU is equally simple: Account energy consumption to the active energy container, or, if there is no active energy container, distribute it evenly among all normal energy containers in the system.

6.2 Shared Use

If the startup overhead of a hardware component is not negligible, other apportioning policies must be used. A policy suited for most devices is to share the collaterally consumed energy among all containers which were associated in the time interval between startup and shutdown of a hardware component. The collaterally consumed energy could either be apportioned evenly to these containers, or proportionally to their hardware usage. We use an evenly apportioning policy for the magnetometer sensor on the MTS300 sensor board, which has a large startup overhead (waiting 100 ms for the sensor to stabilize) and negligible use costs (taking an A/D converter sample is done in a few clock cycles of the MCU). The implementation is integrated into the ICEM [4] framework for shared devices in TinyOS, so that other devices may easily be instrumented as well.

We use the same policy to account the energy consumption of the radio chip in the "low-power listening" mode offered by TinyOS. In this mode, TinyOS repeats the transmission over a configurable time interval, until it either receives an acknowledgment, or a timeout occurs. A node that should receive messages can thus settle on periodically checking for transmissions and keeping the radio chip turned off between checks. The repeated attempts at sending a message can be viewed as a form of synchronization: Barring radio noise, if more messages are sent to the same receiver immediately after one transmission attempt succeeded, those messages will arrive on their first transmission attempt. We treat all transmission attempts but one (the successful one) as synchronization overhead to be accounted to all energy containers of successive messages to the same receiver.

6.3 Continuous Use

Yet another different policy is needed for the radio chip if the application is not configured to use energy-saving mechanisms such as low-power listening. In this case, TinyOS keeps the radio powered on continuously. The absence of use intervals makes it difficult to assign a fair share of collaterally consumed energy to a container. We employ a log of all energy containers that were used to send or receive messages, and apportion collaterally consumed energy of the radio chip to all these containers using a geometric distribution, so that containers using the radio more often will bear most of the energy consumption.

7 Evaluation

We evaluated our energy container system using TinyDB. As a first step, we ported TinyDB to TinyOS 2.1.0. TinyDB is a large sensor-net application consisting of over 140 files with a total of over 25,000 lines. It does not fit in the program memory of a TelosB node (48 kBytes) and uses nearly all program memory of a MICAz node ($\sim$ 60 of 64 kBytes), even with several features such as query sharing and "fancy" aggregations deactivated. The output file of the nesC compiler comprises nearly 40,000 lines of code when TinyDB is compiled for MICAz nodes.

TinyDB is a dynamic sensor-net application in that it allows users to inject queries at run-time, and allows to run a limited number of different queries simultaneously. This makes it an ideal application to benefit from our flexible online energy accounting system.

We evaluated our system with regard to the following aspects:

- Ease of use: The work required to add energy containers to TinyDB.
- Overhead: The additional costs of using energy containers.
- Accounting fairness: Fairness of energy consumption distribution.
- Accuracy: Accuracy of the energy estimation system.

7.1 Experimental Setup

In our evaluation we used two TinyDB applications: *TinyDB-noec* is a regular TinyDB application.

In *TinyDB-full,* which is based on TinyDB-noec, we create an energy container for each new query, and send the energy consumption information in this container back to the base station.

To measure the estimation error of our energy estimation system, we additionally modified TinyDB-full to include a new field in status messages. In this field TinyDB-full reports the difference of the current root container contents to its contents when the first query injection message arrived. Immediately after terminating the last active query on our measured sensor node, we sent a status request message and recorded the energy reported in the status message. Differences between the reported energy values and the measured energy consumption are caused by errors in the energy estimation system.

We used three queries that exhibit different hardware usage. Each of these queries is periodically processed by TinyDB in so-called *epochs*, each epoch being about 750 ms in length by default. At the begin of an epoch, result values are computed for each query, and at the beginning of the next epoch, they are sent out in a query result message. The queries run until they are stopped by a user.

One query, `select nodeid, qids`, uses only information already present in the microcontroller, namely the ID of the node and the IDs of the currently active queries. We used two versions of this query, one using default settings (`sample period 1024`) and one having an epoch length of double the default value (`sample period 2048`).

The third query used, `select nodeid, mag_x`, samples the x-direction of the magnetometer on a MTS300 board, which makes this query consume significantly more energy than the first one.

7.2 Ease of Use

To provide energy containers in TinyDB-full, we had to add 59 lines of code and to make small changes to 5 lines of code. About half of these changes were straightforward changes, like adding fields to message structures and filling them.

7.3 Overhead

We measured two kinds of overhead in our test application: One is the increased code size and memory usage, the other one is additional energy consumption.

As Table 1 shows, adding energy containers to TinyDB caused close to 4000 lines of code to be included in the C file generated by the nesC compiler (which contains the whole application).

Table 1. Sizes of the applications used in our evaluation. Lines of (C) code as reported by cloc (`cloc.sourceforge.net`), Program size and Memory usage as reported by the TinyOS build system.

Application	Lines of code	Program size [byte]	Memory usage [byte]	Avg. current draw [mA]
TinyDB-noec	39175	57382	3292	23.375
TinyDB-full	42971	63552	3449	23.312

We also measured the energy consumption overhead caused by our energy container system. To this end we ran one query (`select nodeid, mag_x`) for about 40 seconds on each of our applications multiple times and measured the current draw. The average current draw is also shown in Table 1. The difference in current draws is 63.1 µA, which is only slightly larger than the standard deviation of the average current draws (which was 31.1 µA for TinyDB-noec and 45 µA for TinyDB-full).

7.4 Accounting Fairness

As an example of how energy containers could be used, we issued two queries with different hardware usage: Both queries requests only information about the software, which is available at virtually no cost (`select nodeid, qids`), but at different sample rates. Query 2 (`sample period 1024`) should send at double the rate of Query 1 (`sample period 2048`). Query 2 is injected after Query 1 and stopped before Query 1, so that energy is accounted first to one, then two, and again one container.

When both queries are active and synchronized, the radio should be used alternately by one and two queries. We configured the sensor node to use the low-power listening mode of TinyOS, and used a shared policy to account collaterally consumed energy on the two energy containers of the queries.

The energy container contents of the queries are reported in the query result messages. Figure 5 shows these energy values plotted as they are sampled at the sensor node. Also shown in the figure is the sum of the most recent energy values of both queries, which should closely resemble the measured energy consumption.

Figure 5 shows that query 2 draws more power than query 1, which can be explained by its higher message sending rate. Query 1 profits from Query 2 in that it is charged with less energy consumption when Query 2 is active.

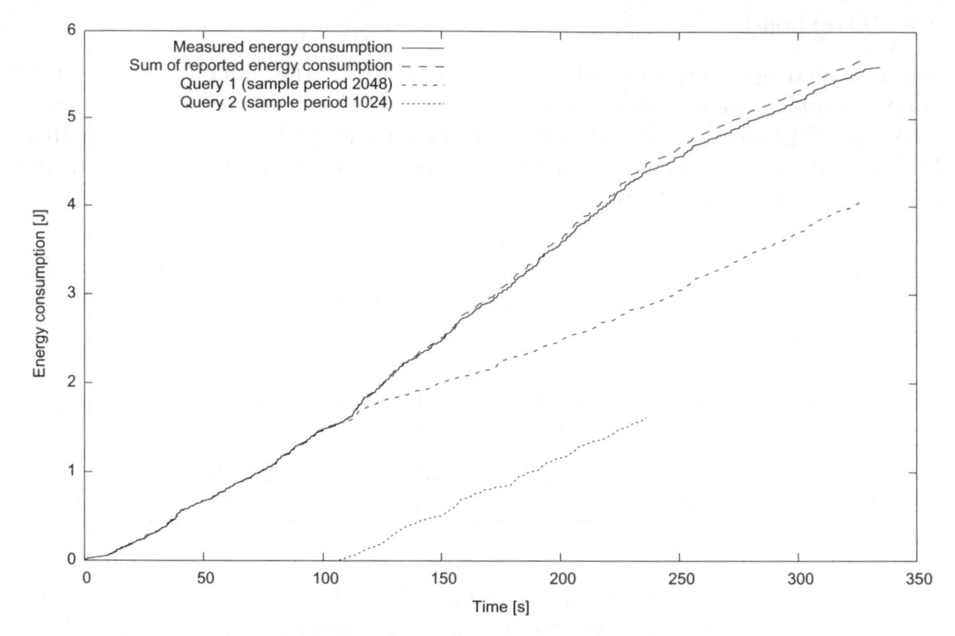

Fig. 5. Energy consumption reported by queries

7.5 Accuracy

To determine the accuracy of our energy estimation system, we measured the real energy consumption of our node and compared the measurements to the contents of the node's root energy container in all of the tests involving TinyDB-full, i.e., some of the overhead tests and the previous example.

The energy consumption recorded in the root container was within 3 % of the measured energy consumption.

8 Future Work

In further work, we plan to improve our implementation to support a greater variety of hardware. Preliminary measurements indicate that the supply voltage has an effect on current draw that varies between chips. We are looking on how best to capture this behavior appropriately in our energy model.

We also plan to incorporate distributed energy management into TinyDB that makes use of our energy-container system.

9 Conclusion

In this paper, we described a flexible online energy accounting system for TinyOS, the basis of which is an online energy estimation system. We introduced energy containers in TinyOS as specialized resource containers, allowing us to account energy consumption of parts of a sensor-net application separately. Evaluation of our implementation shows it to be accurate and to have a low energy overhead.

References

1. Banga, G., Druschel, P., Mogul, J.: Resource containers: A new facility for resource management in server systems. In: Proceedings of the Third Symposium on Operating System Design and Implementation (OSDI 1999), pp. 45–58 (February1999), http://www.cs.rice.edu/~druschel/osdi99rc.ps.gz
2. Dunkels, A., Österlind, F., Tsiftes, N., He, Z.: Software-based on-line energy estimation for sensor nodes. In: Proceedings of the 4th workshop on Embedded networked sensors (EMNETS 2007), pp. 28–32. ACM, New York (2007)
3. Fonseca, R., Dutta, P., Levis, P., Stoica, I.: Quanto: Tracking energy in networked embedded systems. In: Proceedings of the 8th USENIX Symposium on Operating System Design and Implementation (OSDI 2008), pp. 323–338. USENIX Association (December 2008), http://www.usenix.org/events/osdi08/tech/full_papers/fonseca/fonseca.pdf
4. Klues, K., Handziski, V., Lu, C., Wolisz, A., Culler, D., Gay, D., Levis, P.: Integrating concurrency control and energy management in device drivers. In: Proceedings of the twenty-first ACM SIGOPS Symposium on Operating Systems Principles (SOSP 2007), pp. 251–264. ACM, New York (2007)
5. Landsiedel, O., Wehrle, K., Götz, S.: Accurate prediction of power consumption in sensor networks. In: Proceedings of the second IEEE Workshop on Embedded Networked Sensors (EmNetS-II), pp. 37–44 (May 2005)
6. Schmidt, D., Krämer, M., Kuhn, T., Wehn, N.: Energy modelling in sensor networks. Advances in Radio Science 5, 347–351 (2007), http://www.adv-radio-sci.net/5/347/2007/ars-5-347-2007.pdf
7. Shnayder, V., Hempstead, M., Chen, B., Werner-Allen, G., Welsh, M.: Simulating the power consumption of large-scale sensor network applications. In: Proceedings of the 2nd International Conference on Embedded Networked Sensor Systems, SenSys 2004, pp. 188–200. ACM Press, New York (2004)
8. Titzeri, B.L., Lee, K.D., Palsberg, J.: Avrora: scalable sensor network simulation with precise timing. In: Proceedings of the 4th International Symposium on Information Processing in Sensor Networks, IPSN 2005, p. 67. IEEE Press, Piscataway (2005)

TikiriDev: A UNIX-Like Device Abstraction for Contiki

Kasun Hewage, Chamath Keppitiyagama, and Kenneth Thilakarathna

University of Colombo School of Computing, Sri Lanka
{kch,chamath,kmt}@ucsc.cmb.ac.lk

Abstract. Wireless sensor network(WSN) operating systems have resource constrained environments. Therefore, the operating systems that are used are simple and have limited and dedicated functionalities. An application programmer familiar with a UNIX-like operating system has to put a considerable effort to be familiarized with WSN operating systems' Application Programming Interface(API). Even though, UNIX-like operating systems may not be the correct choice for WSNs, some of their powerful, yet simple abstractions such as file system abstraction can be used to overcome this issue.

In this paper, we discuss a UNIX-like file system abstraction for Contiki. File system abstraction is not the panacea. However, it adds to the repertoire of abstractions provided by the Contiki, thus easing the task of the application programmers.

Keywords: Sensor Networks, Device Abstractions, File Systems.

1 Introduction

With the advancements of WSNs, several operating systems have been invented with different features to make the programming easier. Popular WSN operating systems such as TinyOS [1], Contiki [2], SensOS [3] and MantisOS [4] provide location and sensor type dependent access methods. Several concepts such as treating the WSN as a database [5] and a file system [6] have been proposed over the years to ease the application development for WSNs. Other approaches such as SensOS, Contiki and MantisOS provide access to only locally attached devices.

Developing applications for WSN operating systems is a challenging task when compared to general purpose operating systems. One of the main reasons is that the lack of familiar abstractions in WSN operating systems. We observed that an application programmer who is familiar with UNIX-like operating system has to put a considerable effort to be familiarized WSN operating systems' API. In UNIX-like operating systems, devices are accessed as files. It has proven to be a simple, yet powerful abstraction. While UNIX-like operating systems may not be the correct choice for WSNs, some of their powerful abstraction concepts can be incorporated into popular WSN operating systems to overcome above mentioned issue.

P.J. Marron et al. (Eds.): REALWSN 2010, LNCS 6511, pp. 74–81, 2010.

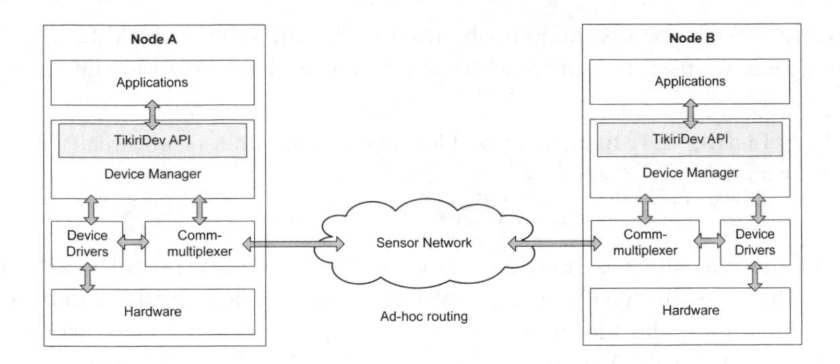

Fig. 1. Communication between two nodes

In this paper, we present the design and implementation of an extension to Contiki operating system which we named TikiriDev. The extension allows application programmers to access local and remote devices in a WSN as files which are named under a single network wide namespace. To be precise, every sensor node in a particular WSN see the same file system abstraction regardless of the application that they use. API calls that are similar to file handling system calls in UNIX-like operating systems such as *open()*, *read()*, *write()*, *close()* and *ioctl()* are used to manipulate the devices of a particular WSN in a location and network transparent manner. In addition to the sensors, nodes may consist of several other components such as actuators and storage media.

2 Design and Implementation

Since TikiriDev is an extension for Contiki to access devices, it is essential to explore the existing device accessing methods in Contiki. Contiki uses three functions, *status()*, *configure()* and *value()* to access sensor devices. These functions have to be implemented in the device driver of each sensor device. In addition, an event based mechanism is also used to notify the availability of asynchronous data. Event is posted to running processes according to the device driver implementation whenever asynchronous data are available.

TikiriDev is composed of three main components: Device Manager, Comm-multiplexer and Device Drivers. For illustration purposes, we have generalized the entire WSN into a network consisting of two nodes(figure 1).

2.1 Device Manager

In TikiriDev, Device Manager provides the illusion of a file system by hiding the underlying complexity. The devices seen by the applications are mapped to the real devices through TikiriDev. Device Manager provides five API calls, *td_open()*, *td_pread()*, *td_pwrite()*, *td_ioctl()* and *td_close()* to access these devices.

TikiriDev API. Typically, UNIX file handling system calls block the calling threads until the requested resource is available. Applications in Contiki are implemented as processes called protothreads. To implement blocking calls inside protothreads, we used protothread spawning method as shown in the listing 1.1.

Listing 1.1. Implementing blocking calls inside a protothread

```
#define td_open(name, flags, fd)  \
        PROCESS_PT_SPAWN(&((fd)->pt), \
                   td_open_thread(name, flags, fd))
```

Since a function which is used as a protothread has only protothread specific return values, as in UNIX *open()* system call, file descriptor cannot be returned. Therefore, the file descriptor is given as a reference type argument to the *td_open()* API call. The prototypes of the five TikiriDev API calls are shown in the listing 1.2.

Listing 1.2. The prototypes of TikiriDev API

```
td_open(char *name, int flags, fd_t *fd);
td_close(fd_t *fd);
td_pread(fd_t *fd, char *buffer, unsigned int count,
                                 unsigned int offset, int *r);
td_pwrite(fd_t *fd, char *buffer, unsigned int count,
                                  unsigned int offset, int *r);
td_ioctl(fd_t *fd, int request, void *argument, int *r);
```

The API calls *td_pread()* and *td_pwrite()* are analogous to the UNIX *read()* and *write()* system calls respectively. However, TikirDev API calls take an extra argument `offset`. The argument `offset` provides adequate amount of information for the device drivers to perform an operation like *lseek()* itself. Therefore, we do not provide specific API call to increase/decrease the file pointer when accessing a storage medium. When reading a device like a temperature sensor which has an unbounded data stream, the argument `offset` is discarded.

The API call *td_ioctl()* is used to configure the devices. Moreover, this API can also be used to receive notifications when asynchronous data is available.

Device tables. Two tables are used to keep the information about the devices. Local Device Table is used to keep the information about locally attached devices whereas Remote Device Table is used for the devices on remote nodes. Figure 2 shows the structure of both tables.

File Descriptors. In UNIX-like operating systems, there is a file descriptor table per process. In TikiriDev, instead of a per process file descriptor table, a global table is being used. Further, instead of using integer type file descriptors used in UNIX-like systems, we defined a C structure as the file descriptor as shown in the listing 1.3.

Since blocking functions in Contiki are protothreads, we have to use it to implement blocking context in our API calls. TikiriDev transparently spawns a protothread when an application calls its API calls. Therefore, we embodied the required control structure into the file descriptor itself. However, the member variable `fd` can be used similarly as UNIX file descriptors.

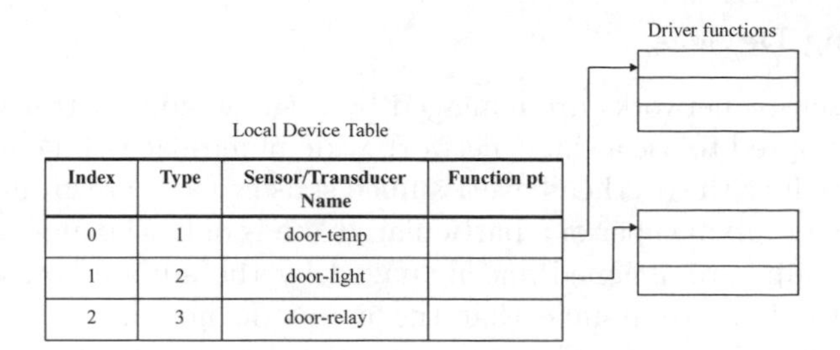

Remote Device Table

Index	Type	Sensor/Transducer Name	Remote index	Node address
0	1	/room/door-temp	0	10.21
1	2	/room/door-light	1	10.02
2	3	/kitchen/door-relay	2	05.07
4	3	/kitchen/fan-relay	2	05.10

Fig. 2. Local and Remote device tables

Listing 1.3. The C structure which is used as the file descriptor

```
typedef struct fd {
  struct pt pt; /* handler for the newly spawned child protothread */
  int fd;       /* Actual file descriptor. */
} fd_t;
```

2.2 Comm-multiplexer

Comm-multiplexer handles all inter-node communications related to device manipulation. This component is built on top of Rime [7] communication stack which is the default radio communication stack of Contiki. Since this component handles multiple device requests from Device Manager, we extended some of the Rime's communication primitives to multiplex several requests via same connection instead of having a connection per request.

2.3 Device Drivers

Device Manager does not access devices directly. Instead it uses device drivers to access devices. For any device, the device driver can be represented using a simple data structure as listed in the listing 1.4. Device driver provider must implement five driver functions which correspond to the API calls *td_open()*, *td_close()*, *td_read()*, *td_write()* and *td_ioctl()*.

Listing 1.4. The device driver representation in TikiriDev

```
struct dev_driver {
  int (* init)(void);
  int (* read)(void *buf, unsigned int count, unsigned int offset);
  int (* write)(void *buf, unsigned int count, unsigned int offset);
  int (* ioctl)(int request, void *data);
  int (* close)(void);
};
```

2.4 Naming Devices

Most of the sensor networks are managed by a single administrative authority. Therefore, we opted to use a single network wide namespace to name the devices of nodes. To reduce the overhead, we assumed sensor nodes are grouped in such a way that each group member of a particular group is only a single hop away from others and groups are assigned and managed by the administrative authority. In addition to that, we assume that the nodes do not move. That is, group membership of a device does not change over the time unless it is explicitly altered by the administrator.

In TikiriDev, there is a root folder which is denoted by the symbol "/". Each group is represented as a sub folder under the root folder. All the devices in the network are represented as files under the sub folders of the root folder. For an example, "*/room1/door-temp*" represents the temperature sensor mounted on some node located in room 1.

It is not possible to share a name between two members of the same group(i.e: There cannot be two files with the same name in a particular folder). In the same way, two folders also cannot have the same name. This implies two groups cannot share the same name. Any application running on any sensor node sees the same file system abstraction as described above.

3 Evaluation

To test the implementation of TikiriDev, we used Scatterweb MSB430 sensor node platform and Sky sensor node simulation of Cooja simulator [8].

The cost of accessing remote devices is higher than that of local devices due to radio communication. Table 1 shows the additional data transmitted during each API call when accessing a remote device. The API call *td_open()* has the biggest data overhead when compared to other API calls since the group name and the device name are included in device discovery request messages.

Table 1. Additional data transmitted during the each API call when accessing a remote device

API call	Size(bytes)
td_open()	31
td_pread()	8
td_pwrite()	8
td_ioctl()	10

As shown in Table 2, the code size of TikiriDev (program memory usage) is just 24384 bytes and RAM usage is 3307 bytes.

We also measured stack usage of TikiriDev (including Contiki) when accessing local and remote devices. To measure the stack usage, we implemented an application that writes/reads LED on/off state to/from local/remote node in response to a button click event.

Table 2. Memory footprint of TikiriDev with Contiki OS in bytes for MSB430 platform

	Program memory	Static memory
Contiki OS(without TikiriDev)	18880	2579
TikiriDev with Contiki OS	24384	3307

The stack memory usage for that application when accessing the local LED is shown in the figure 3. The spike shown in the figure 3 indicates the maximum stack memory usage. According to the test results, the maximum stack memory usage is about 140 bytes. The figure 4 shows the stack memory usage of a node when accessing a LED on a remote node while figure 5 shows the stack memory usage of the node whose device is being accessed.

The maximum stack memory reported during a remote LED access is less than 250 bytes. Therefore, the maximum memory consumed is summed up to 3557 bytes. Though this is a significant amount of memory for a MSB430 node which has a RAM of 5 kilobytes , it is an insignificant amount for a node similar to Sky which has a RAM of 10 kilobytes.

The syntax of TikiriDev API is not mapped directly to the syntax of the UNIX file handling system calls. The main reason behind that is the implementation differences of the blocking function calls in Contiki and UNIX-like operating systems.

4 Related Work

Plan 9 was the first distributed operating system that used a file centric approach to view the entire system [9]. In Plan 9, the application's view of the network is a single, coherent namespace that appears as a hierarchical file system but may represent local or remote resources. FISN [10] and the file system abstraction proposed by Tilak et al. [11] are two approaches inspired by Plan 9 and use file system servers to map the WSN into files. LiteOS is another file system based approach which provides a wireless shell interface to interact with the WSN using

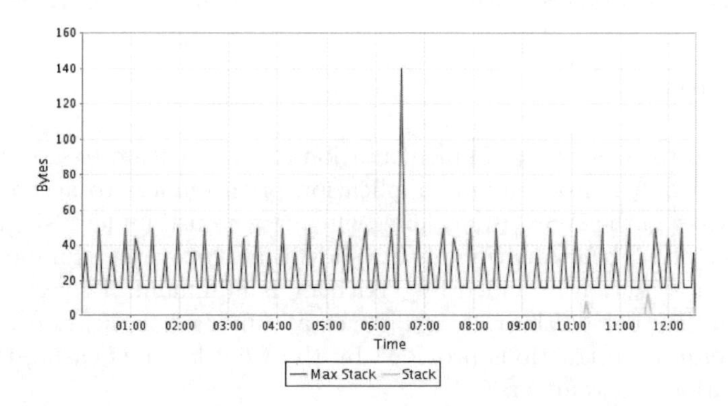

Fig. 3. Stack memory usage when accessing the local LED

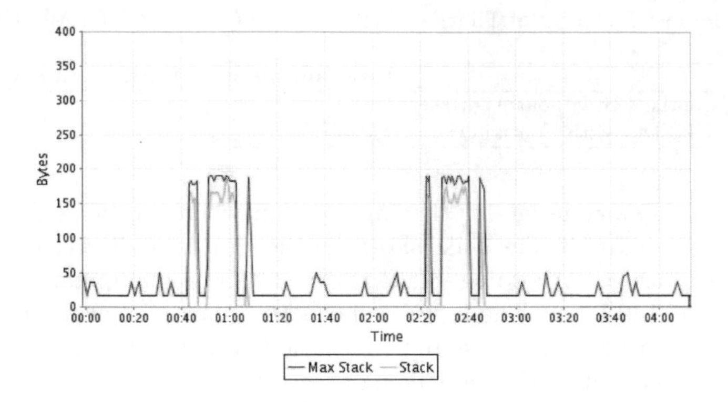

Fig. 4. Stack memory usage of the node in which the application runs

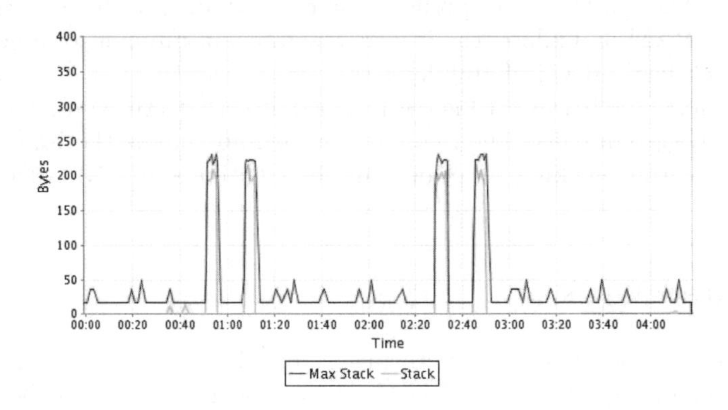

Fig. 5. Stack memory usage of the remote node whose LED is being accessed

UNIX style commands [6]. These approaches provide the file system abstraction only at a special (usually a resource rich) node connected to the WSN. SensOS [3] and MantisOS [4] provide access to only locally attached devices. However, the file system abstraction provided by TikiriDev is available on all the sensor nodes.

5 Conclusion

We presented the design and implementation of an extension to Contiki operating system. The extension allows application programmers to access local and remote devices in the WSN as local files which are named under a single network wide namespace. The evaluation results show that the overhead imposed by the file system abstraction provided by TikiriDev is minimum. Though file system abstraction may not be the right choice for all kinds of applications, it adds to the repertoire of abstractions provided by the Contiki, thus easing the task of the application programmers.

References

1. Hill, J., Szewczyk, R., Woo, A., Hollar, S., Culler, D., Pister, K.: System architecture directions for networked sensors. In: Architectural Support for Programming Languages and Operating Systems, pp. 93–104 (2000)
2. Dunkels, A., Grïnvall, B., Voigt, T.: Contiki - a lightweight and flexible operating system for tiny networked sensors. In: Proceedings of the First IEEE Workshop on Embedded Networked Sensors (Emnets-I), Tampa, Florida, USA (November 2004)
3. Yang, M., So, S.S., Eun, S., Kim, B., Kim, J.: Sensos: A sensor node operating system with a device management scheme for sensor nodes. In: Proceedings of the International Conference on Information Technology, pp. 134–139 (2007)
4. Bhatti, S., Carlson, J., Dai, H., Deng, J., Rose, J., Sheth, A., Shucker, B., Gruenwald, C., Torgerson, A., Han, R.: Mantis os: An embedded multithreaded operating system for wireless micro sensor platforms. In: ACM/Kluwer Mobile Networks and Applications (MONET), Special Issue on Wireless Sensor Networks, pp. 263–279 (2005)
5. Madden, S., Franklin, M.J., Hellerstein, J.M., Hong, W.: Tinydb: An acquisitional query processing system for sensor networks. ACM Trans. Database Syst. 30(1), 122–173 (2005)
6. Cao, Q., Abdelzaher, T., Stankovic, J., He, T.: The liteos operating system: Towards unix-like abstractions for wireless sensor networks. In: Proceedings of the 7th International Conference on Information Processing in Sensor Networks (ACM/IEEE IPSN), pp. 233–244 (2008)
7. Dunkels, A.: Rime — a lightweight layered communication stack for sensor networks. In: Proceedings of the European Conference on Wireless Sensor Networks (EWSN), Poster/Demo session, Delft, The Netherlands (January 2007)
8. Osterlind, F., Dunkels, A., Eriksson, J., Finne, N., Voigt, T.: Cross-level simulation in cooja. In: Proceedings of the European Conference on Wireless Sensor Networks (EWSN), Delft, The Netherlands (January 2007)
9. B, H.: Reinventing unix: an introduction to the plan 9 operating system. Library Hi Tech
10. Horey, J., Tournier, J.C., Widener, P., Maccabe, A.B., Kilzer, A.: A filesystem interface for sensor networks. Technical report, Department of Computer Science in University of New Mexico and Department of Computer in Science Gonzaga University (2008)
11. Tilak, S., Pisupati, B., Chiu, K., Brown, G., Abu-Ghazaleh, N.: A file system abstraction for sense and respond systems. In: Proceedings of the 2005 Workshop on End-to-end, Sense-and-Respond Systems, Applications and Services, pp. 1–6 (2005)

Location Based Wireless Sensor Services in Life Science Automation

Benjamin Wagner[1], Philipp Gorski[1], Frank Golatowski[2], Ralf Behnke[1], Dirk Timmermann[1], and Kerstin Thurow[2]

[1] Institute of Applied Microelectronics and Computer Engineering
University of Rostock, Germany
{benjamin.wagner,philipp.gorski,dirk.timmermann}@uni-rostock.de
[2] Center for Life Science Automation, CELISCA
Rostock, Germany
{frank.golatowski,kerstin.thurow}@celisca.de

Abstract. Over the last years Wireless Sensor Networks (WSN) have been becoming increasingly applicable for real world scenarios and now production ready solutions are available. In the same period the upcoming combination of Service-oriented Architectures and Web Service technology demonstrated a way to realize open standardized, flexible, service component based, loosely coupled and interoperable cross domain enterprise software solutions. But those solutions have been too resource-intensive and complex to be applicable for limited devices like wireless sensor nodes or small-sized embedded systems. Thus, more and more research investigations have been launched to bring the aspect of cross domain interoperability to the field of embedded battery powered devices. The proposed laboratory assistance solution in this paper demonstrates the benefits of Web Service enabled WSNs for process monitoring and disaster management by extending an existing system in the Life Science Automation domain. Especially, the capability to provide location based services in industrial automation environment represents a beneficial feature of the presented integration approach and results in high-quality information delivery bundled with specific data about the locational origin of the capturing sensor.

Keywords: Devices Profile for Web Services (DPWS), Disaster Management, Laboratory Information Management System, Life Science Automation (LSA), Sensor Web Enablement (SWE), Sensor Observation Service (SOS), Service-oriented Architecture (SOA), Web Services.

1 Introduction

In the domain of Life Science Automation (LSA) the experimental setups, laboratories and appliances, needed for complex automated chemical and/or biological screening analysis typically consists of closed proprietary and highly specialized solutions. These solutions are mainly configured to obtain a high throughput and satisfy the required environmental constraints for the experiments. Such closed process chains are especially

P.J. Marron et al. (Eds.): REALWSN 2010, LNCS 6511, pp. 82–93, 2010.

designed to solve characteristic classes of analysis problems with high efficiency. The available integrated sensors/actors are wired and have fixed positions at the laboratory appliances. This makes it a hard challenge to extend or adapt them dynamically without high efforts. Using wireless sensors instead offers the needed flexibility. The additional equipment of localization capabilities for those wireless sensor nodes will enable locational tagged measurement data in combination with sensing/acting services. Furthermore, these capabilities will provide an easy reconfiguration of wireless sensor network appliances. However, today´s wireless sensor nodes are not equipped with sensors needed in life sciences, e.g. CO and H_2. In this paper we introduce wireless sensor nodes addressing the needs of life sciences.

Another important challenge relies on the handling and usability of those appliances for non-technical employees. The typical scientist, who utilizes the laboratory for the experiments, is not skilled enough on the technical domain to setup, reconfigure, maintain or extend the existing workflow.

The integration of Wireless Sensor Networks (WSNs) in enterprise systems will be the right way for the future to realize a flexible and extendable system to overcome these above mentioned deficiencies. Especially, those enterprise solutions with a tight coupling of higher software layers for enterprise process management and data processing to the underlying control of industrial production processes can benefit from the flexible integration of WSNs. Web Service technology has a great potential to support a seamless integration of WSNs and to achieve an advanced applicability. Moreover, especially the benefits for usability and abstraction of a technical system providing services will be focused by our presented work.

An existing web based Laboratory Information Management System (LIMS) at the CELISCA laboratories and the corresponding appliances are extended with the WSN (see Fig. 1). The LIMS maps the workflow of an existing analysis process chain and supports the evaluation of data measured. We will show, that the WSN based service infrastructure, developed in this work, increases the flexibility, extensibility and usability of a given wired laboratory setup. Furthermore, we provide new interaction concepts for system control, setup and configuration.

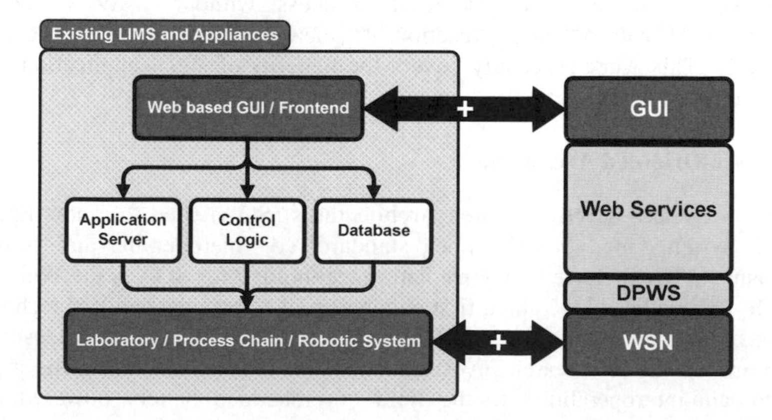

Fig. 1. Integration points of the Wireless Sensor Network into the existing Laboratory Information Management System at the CELISCA laboratories

This paper is organized as follows. In section 2 we briefly review WSN technology, service-infrastructures, and technologies for disaster management. In section 3 we give an overview of the developed system, based on Devices Profile for Web Services (DPWS) and Sensor Web Enablement (SWE) middleware, to realize disaster prevention system for LSA. In Section 4 we describe user interaction with wireless sensor network and section 5 provides concrete application scenarios for our solution. We emphasize our Laboratory Assistance WSN and its innovations, at different levels of abstraction with a detailed overview, due to our integration and deployment concepts and details about the hardware of WSN. Finally the article ends with a conclusion.

2 State of the Art

2.1 Wireless Sensor Networks

Recent technological advances enabled the development of tiny wireless devices which are referred to as Wireless Sensor Nodes. Those devices usually consist of a number of physical sensors, gathering environmental data like temperature or light, a microcontroller, processing the data, and a radio interface to communicate with other nodes. These devices are typically battery driven to allow autonomous work and wireless deployment. Wireless Sensor Networks are interconnected assemblies of such devices [1]. In recent years, much work has been done on the various aspects of the wireless sensor networks, especially on the communication level and has result in standardized communication interfaces, like ZigBee, Bluetooth Low Energy, 6LoWPAN, Wireless HART, and SP100. However, developing wireless sensor nodes with low powered sensors measuring typical gases, which are used in life sciences, is still a challenging task. We have developed wireless sensor nodes which can measure carbon monoxide (CO) and hydrogen (H_2), two very dangerous gases. In the industrial domains WSNs become more and more attractive due to its flexibility, sizing dimensions and ease of use [2,3,4]. The main focus relies on process control and monitoring applications. In contrast to traditional wired sensors, WSN nodes can be easily placed, as close as possible to the process, without costly wiring and in combining those with actuators, reactions to measurement events can be initiated immediately. This work especially covers the benefits of WSN applications for the industrial domain of LSA.

2.2 Service-Oriented Architecture

Over the last years, Service-oriented Architectures (SOA) tried to renew Enterprise Software Systems in a flexible, open standardized, interoperable and component based manner. The preferred implementation technology for SOA is the Web Service approach. But the heavy weighted first generation of upcoming standard technologies were not suitable for mobile and limited embedded devices like wireless sensor nodes. Thus, more and more research investigations were launched to bring the aspect of cross domain interoperability to the field of embedded battery powered devices [5,6,7]. The results were combined in the Devices Profile for Web Services (DPWS) [8], which represents the official OASIS standard for the seamless integration of embedded mobile systems into the Web Service concepts. In the domain of sensor applications the Open Geospatial Consortium (OGC) founded the initiative for Sensor Web Enablement (SWE) and released a collection of open standards [9]. These standards realize the high level management of sensor data and networks, accessible via Web Service technologies. Especially the Sensor Observation Service (SOS) is highly relevant for our research investigations. Our solution builds up on the above mentioned standards to realize the interoperable device connectivity and the management of sensor data.

2.3 Managing Disasters and Incidences

An automated and autonomous solution for the process observation and disaster management represents an essential part in our integration concept for WSNs. While the WSN is responsible for the data delivery another instance has to evaluate measured data and must decide if defined constraints for the processes are met. Further, correct reaction must be initiated to guarantee the behavioral correctness of an observed process. Most commonly disaster management is used as a synonym for emergency management. It deals with natural and human based disasters, like earth quakes or explosions. The four phases of emergency management are 1. *mitigation*, 2. *preparedness*, 3. *response* and 4. *recovery* [10]. The mitigation phase focuses on the prevention of that hazards will become disasters. The preparedness phase contains the development of plans for the treatment of occurring disasters. The response phase includes mobilization and coordination of emergency services, e.g. police and ambulance. Recovery treats restoring of affected areas and infrastructures. An actual example of an emergency management system (EMS) is SAHANA [11]. It impressively shows that a main purpose of EMS is to deal with a kind of resource management, planning and coordination in the case of present disasters. Another project, dealing with that topic, called SoKNOS, is further described in [12].

In contrast to the above mentioned description for disaster management, the solution of this paper focuses on the mitigation phase. The goal is to detect incidences and hazards as soon as possible to prevent disasters in observed LSA environment to ensure a correct analysis procedure. Therefore an observation service has been implemented, which analyses actual sensor readings to initiate alarm chains and react on abnormal environmental parameters.

2.4 Localization Systems

Numerous technologies and methods for locating objects were developed in the past. They differ in accuracy and reliability of the measurements, and the susceptibility to other systems or physical obstructions. Localization algorithms can be classified by their use of different parameter as inputdata for their calculation of sensor node locations [14]. The inputdata will be approximated parameters like geometric distances, locational angles and areas, topological hop counts and neighbourhood relations between sensor nodes. Furthermore, a various number of different methods like the measurement of received signal strengths (RSS), time difference of arrival (TDoA) or angle of arrival (AoA) enables the estimation of those needed parameters to calculate the sensor location [15]. Because there is often a tradeoff between the accuracy of location approximations and the energy consumption the algorithms need for the parameter extraction, localization methods can be classified into coarse-grained and fine-grained [14]. For the presented approach we decided to use the commercial Ubisense system. This is a real-time localization system based on UWB radio technology, and especially delivers the needed accuracy for indoor localization [16].

3 System Overview

The following section includes an overview of our WSN infrastructure, summarizes the basic architecture and gives detailed descriptions of the used hardware, service components, workflow and the WSN itself. The core components of the solution are illustrated in the architecture overview below (see Fig. 2.).

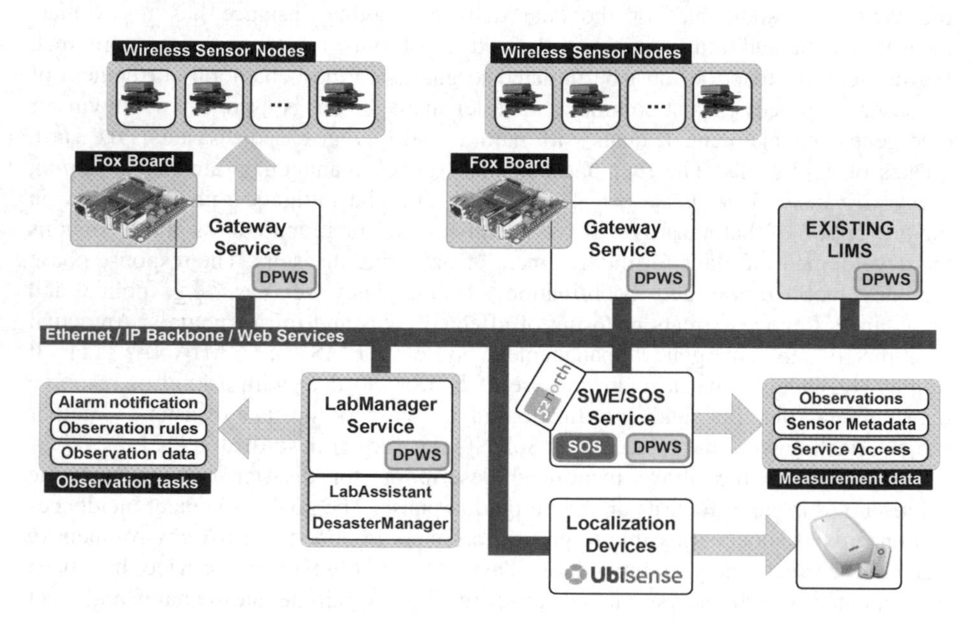

Fig. 2. Schematic overview for the complete system architecture and its components

3.1 Hardware

The required hardware for our WSN infrastructure consists of the wireless sensor nodes itself and a gateway to enable service based interaction with the WSN. Both elements are optimized for energy aware processing.

Wireless Sensor Nodes. The wireless sensor nodes have to fit requirements like high robustness, autonomous acting, small sizing dimensions and a long battery life. To meet these requirements we build up a WSN node based on the eZ430-RF2480 platform for wireless communication via ZigBee technology, from Texas Instrument and extended this platform with the needed modules for additional measuring, communication and energy supply capabilities. To meet the special requirements in

LSA, the needed sensor add-ons for measuring phenomena like temperature, gas concentrations (CO and H_2), light intensity, battery voltage and vibration were integrated. Thus, the WSN node can be integrated in typical data capturing scenarios of the LSA domain. The gas sensors are based on the electro-chemical measurement principal. This avoids the necessity for active heating and only a small current is needed, depending on the gas concentration. The figure below shows the resulting WSN node in combination with different sensor modules (see Fig. 3).

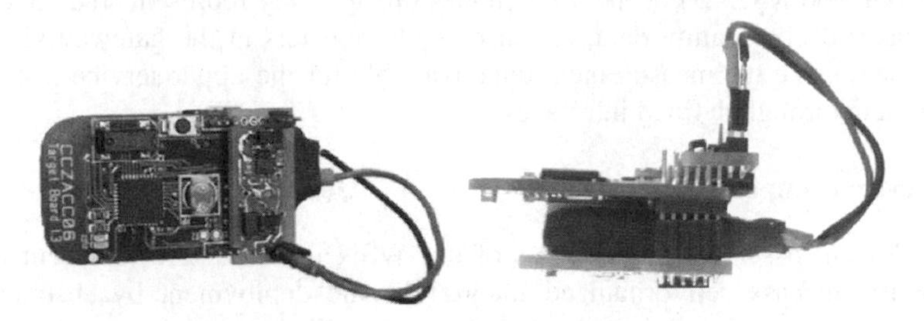

Fig. 3. Illustration of the customized wireless sensor node used at the CELISCA labs

This WSN node has a sizing dimension of 38 x 38 x 48 mm³. The energy source of the node consists of an internal lithium-polymer accumulator with a capacity of 100 mAh. Through an integrated mini-USB connection this accumulator can be recharged. To configure or update the software of a wireless sensor node, the ZigBee or the mini-USB interface can be utilized. For this purposed WSN node we have evaluated the runtime behavior and the energy profile to enable an optimal sensor lifetime prediction and adaption for experimental setups. Thus, the sensor nodes can be configured to fit the needs of an experiment regarding the accruing amounts of measurement data and the expected total runtime. First, the total runtime/lifetime of the WSN node (with acceleration sensor add-on) over a variation of the sampled-data period was evaluated (see Fig. 4 left data plot). The result fully fits the needs of data capturing of the experimental setups at the CELISCA laboratories. Furthermore, the energy profile for the variation of the data transmission period was evaluated for our sensor platform including additional add-ons for CO and acceleration measurements

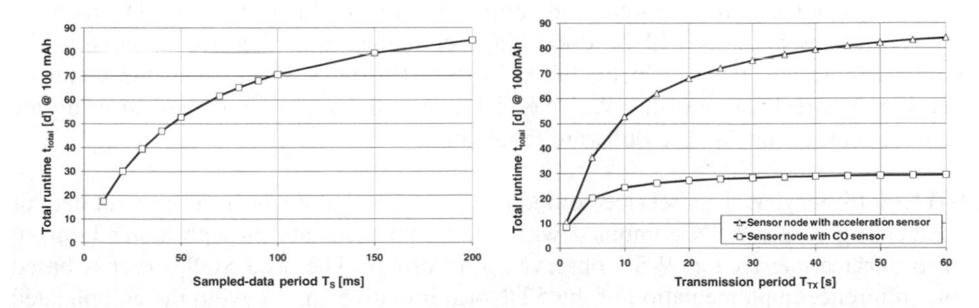

Fig. 4. The left data plot contains the total runtime capability of the customized wireless sensor node over the variation of the sampled-data period. The right data plot illustrates sensor lifetime over the variation of the data transmission period for the two different sensor add-ons.

(see Fig. 4 right data plot). Both data series show that the used sensor platform is best suited for short-term experiments with high data capturing rates and long-term experiments with the need for continuous observations of environmental parameters.

WSN Gateway. The Fox Board LX832,produced by Acme Systems, is a compact embedded Linux server system and represents the service gateway for the WSN nodes in our solution. This board is suited with integrated interfaces for USB 1.1, Ethernet 10/100, IDE and RS232. For the WSN nodes this gateway represents the collector for their measured observation data. A sensor node registers at the gateway via ZigBee and this will make the measurement data available for the upper service instances by serving them through defined interfaces.

3.2 Service Components

The implemented service components of our WSN infrastructure represent the core concept for an easy self organized integration and deployment by abstracting the hardware and connection details through devices services.

LabManager. This service component represents the core of the disaster prevention and the process monitoring. It includes a service for notification events, a DesasterManager for the process observation and the needed functionalities to interact with the other basic system service components. The LabManager runs observation tasks, which were configured via the LabAssistant component, and is able to perform several of those tasks in parallel. An observation task consists of the following configuration subset:

– A set of observations which represents the abstract WSN measurement data and delivers the input parameter for the observation rules.
– A set of observation rules (rule set) which describes system reactions triggered by the input of the WSN observation data.
– A set of notification events (SMS, Mail, Beep or combined alarm action) that will be triggered if a violation of the corresponding observation rule takes place.

The behavior of the LabManager Service is comparable to a specialized workflow engine, which has multiple inputs and controls the firing of alarm events. The requests for sensor observations will be done via a combination of Service Discovery and Publish/Subscribe mechanism, included in the SWE/SOS Service and the Gateway Service. The realization of the Web Service connectivity was implemented using the Axis2 engine of the Apache Software Foundation.

SWE/SOS Service. This service component provides the sensor measurement data of the heterogeneous WSN combined with additional metadata through a standardized Web Service interface as WSN observation offerings. The used SOS server is based on a reference implementation of the 52°North initiative and to avoid the complicated generating/parsing of XML requests, the corresponding OX Framework is used to access and configure the SOS Services (see Fig. 5). The OX Framework offers a simplified access to SOS server via method calls. The input parameters for these methods will be send as JSON formatted set over HTTP. Defining new sensor observations and the SWE/SOS Service configuration will be realized via the SOS Assistant web frontend. The SOS Service includes three core functionalities to request all metadata about the SOS Service and its offerings, the specified observation data and the corresponding metadata, and to get detailed information about the specified sensor, which provides the observation data. Additional transactional operations enable the registration of new sensors and insertion of new observations.

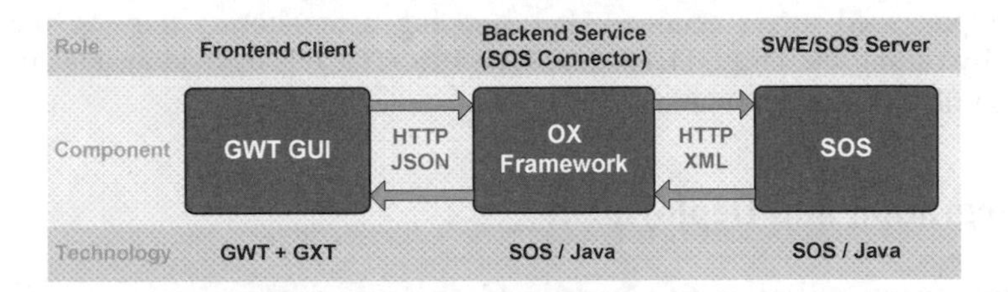

Fig. 5. Exemplary schematic block diagram to illustrate the functionality of the OX Framework in combination with the GWT frontend of the Lab Assistant

Gateway Services. The Fox Board Gateway is abstracted through Web Services, implemented using the DPWS technology stack WS4D-gSOAP [13]. These services enable WSN nodes to dynamically discover and connect to the gateway, forward the WSN observation data to the SWE/SOS Service, discover sensors, register new sensor nodes, request locally stored sensor data and let other services subscribe for defined WSN observation events. Several Fox Board gateways and its services can run in parallel to provide a scalable and reliable access to the WSN, without concurrent behavior.

3.3 LabAssistent and SOS Assistant

The LabAssistant is a web based frontend to configure the WSN, the Gateway and the LabManager Service. It is realized as fat client web application using the AJAX technology of the open source Google Web Toolkit (GWT) and its support for Remote Procedure Calls. Additionally, the assistant guides a user through the integration and deployment process of the WSN. The SOS Assistant is similar to the LabAssistant. It provides a web based frontend to configure the SWE/SOS Service of our solution. This includes the management of sensors, observations and the graphical illustration of observation requests/data.

LabAssistant. This component represents a web application realized with a GWT frontend, which provides the main interface for managing monitoring processes and observation tasks. The setup is divided into two subsequent flows. First of all, the setup of the sensor network in the specific observation environment has to be executed. Afterwards, the rule sets and additional data for the observation tasks have to be defined. The LabAssistant guides the user through this procedure and abstracts the underlying technological processes. Furthermore, there is no need for the user to edit configuration files or other formats, because the LabAssistant will generate them itself in a XML format.

DesasterManager. The DesasterManager is a web service component, which provides the monitoring/observation functionality. It is implemented upon the Axis2 Framework of the Apache Software Foundation. This service component provides four methods with a specific set of parameters to control the integrated multi-threaded rule engine (JRuleEngine). The START-method initializes a monitoring thread. This thread starts a new instance of the rule engine when new measurement events of defined sensor nodes arrive. The STOP-method finishes the defined observation tasks when a running experiment ends. Finally, it generates a summarizing log file that contains all executed/processed server actions. After stopping a monitoring this log file can be deleted by calling the DELETE-method. The PROTOCOL-method enables the access to the log file of currently running or stopped observation tasks.

4 Integration and Deployment

The deployment and integration strategy of our laboratory assistance solution raises the functionality, flexibility and usability of the existing LIMS system to a new level, results in cost efficient workflow turnarounds and reduces setup times. Especially the usability advantage for non-technical skilled users is a real innovation of our solution. The user now handles services of the WSN and is able to place the sensors where he needs them for his experiment. Without our WSN infrastructure the user had to be or to call a specialist, if changes in the appliances of the experiment had to be made. Especially when changing the wired sensors in their positions or measurement services. They have to be rewired, tested or recalibrated, and their new services had to be implemented. With our solution the user is able to change the hardware, replace or relocate sensors, without the need to change the software or anything else, because the setup of an experiment is bound to services of the WSN and not to the sensor hardware itself. This makes it possible to work with components-of-the-shelf WSN nodes suited with defined sensors.

Deployment. Setting up a new WSN deployment for an experiment becomes a simple procedure. Before creating a new experiment, the sensors have to be placed/plugged at the laboratory appliances and the existing sensors must be checked out. The wireless sensor nodes register themselves and their observation offerings at the Gateway Service nearest to them.

Integration. The user initiates a new experiment via the LabAssistant and creates the needed observation tasks for the process monitoring. These tasks will be suited with the necessary observation rules and the corresponding input parameter from the previously installed sensor observations. Each rule violation will be bound to a specific notification. The available observation offerings of the WSN nodes will be discovered automatically by the LabManager and the user has to pick the right ones from a list to assign them to the observation rules. If different sensor data should be combined to new observations, the user is able to create those combinations through the SOS Assistant. Assigning sensor observations to an experiment includes an automated subscription of the LabManager for observation events at the SWE/SOS or Gateway Service. The location assignment for the WSN nodes by the user will be realized through the use of the Ubisense localization tag. This tag will be placed near to the WSN node and over a push button on it the user initiates the position measurement by the wall mounted sensors. Afterwards, the calculated WSN node position will be send to the LabManager service and the user has to assign it to the right WSN node instance in the experimental setup.

After the successful deployment procedure the experiment will be started and runs now with our integrated disaster prevention. While the experiment runs external applications are able to request the corresponding observation time series via the Web Service interface of the SOS server. When the experiment ends the WSN nodes will be picked up by the user and recharged at a charge station. The complete observation data is stored in an SOS server database and will be served through a Web Service for the post data processing. The existing LIMS runs in parallel to our solution and is not affected.

5 Application Scenarios

There exist three main scenarios the proposed solution will be used for at the CELSICA laboratories (see Fig. 6). These scenarios differ in the granularity of objects that have to be observed (rooms, devices or experiments), and in the associativity that the WSN nodes will have to the experiments or how the nodes will be involved in the experimental process flow (static and dynamic conditions). Furthermore, the WSN node can provide actor services to regulate or control the observed environmental conditions.

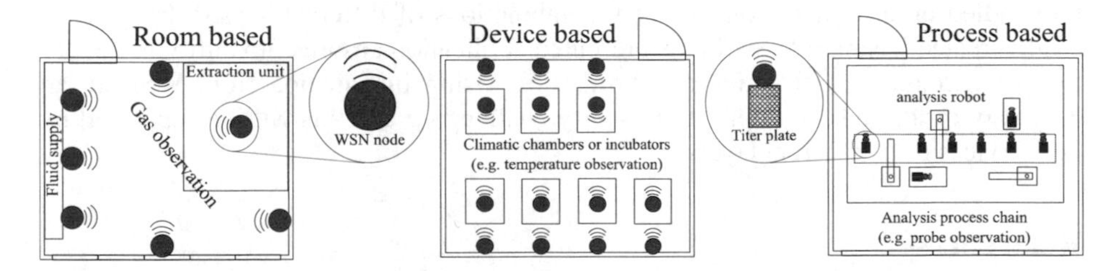

Fig. 6. Three major application scenarios for the proposed solution: room, device and process based observation strategies

5.1 Room Based Monitoring

The room based monitoring focuses the observation of environmental parameters independent of the experimental setups and appliances the room contains. Thus, multiple laboratory environments will be supported and the WSN represents the hazard/risk detection system to prevent dangerous situations regarding the laboratory personal or appliances. At the CELISCA laboratories this scenario is used to observe gas concentrations (H_2 or CO) and the room temperature.

5.2 Device Based Monitoring

The device based monitoring scenario associates WSN nodes directly to single instances of laboratory appliances (climatic chamber or incubators) for specified observation tasks. This scenario introduces the possibility to enhance devices with new sensing/acting services or to refine existing measuring capabilities to achieve a higher measurement accuracy, sampled-data rate or observation density.

5.3 Process Based Monitoring

The third scenario type is represented by the process based monitoring. This observation strategy focuses on the process flow of a single experimental setup. The WSN nodes are tagged directly to single experimental probes (like a titer plate or other samples) without a fixed position and will pass through all experimental stages. Thus, a completely closed observation of the process chain can be guaranteed, and parameter variations or environmental changes for single instances of an experiment can be observed at an early stage.

6 Conclusion and Future Work

In this paper we have presented a SOA to integrate wireless sensor networks into an existing laboratory information management system (LIMS). The architecture uses DPWS based Web Services for the collaboration and orchestration of devices, abstracted as service instances. Thus, a decoupling of the hardware and higher functionalities were reached, with the additional benefit of a higher usability and flexibility. Furthermore, the functionality of SWE/SOS is now available to the Life Science Automation domain, which is very beneficial, because measurement data will be handled on a higher level and new combinations of different sensor data can be easily created. Using a WSN in the presented application ensures flexibility necessary to construct future Life Science Laboratories. Using information from WSN in an easy way inside today's application is very challenging and this will be supported by our services over different layers.

References

1. Yick, J., Mukherjee, B., Ghosal, D.: Wireless sensor network survey. Computer Networks 52, 2292–2330 (2008)
2. Jiang, P., Ren, H., Zhang, L., Wang, Z., Xue, A.: Reliable Application of Wireless Sensor Networks in Industrial Process Control. In: The Sixth World Congress on Intelligent Control and Automation, WCICA 2006, pp. 99–103 (2006)
3. Bonivento, A., Carloni, L., Sangiovanni-Vincentelli, A.: Platform-Based Design of Wireless Sensor Networks for Industrial Applications. In: Proceedings of the Design Automation & Test in Europe Conference, pp. 1–6 (2006)
4. Antoniou, M., Boon, M., Green, P., Green, P., York, T.: Wireless sensor networks for industrial processes. In: 2009 IEEE Sensors Applications Symposium, pp. 13–18 (2009)
5. Jammes, F., Mensch, A., Smit, H.: Service-oriented device communications using the devices profile for web services. In: Proceedings of the 3rd International Workshop on Middleware for Pervasive and ad-hoc Computing, p. 8. ACM, New York (2005)
6. Leguay, J., Lopez-Ramos, M., Jean-Marie, K., Conan, V.: Service oriented architecture for heterogeneous and dynamic sensor networks. In: Proceedings of the Second International Conference on Distributed event-based Systems DEBS 2008, p. 309 (2008)

7. Zeeb, E., Bobek, A., Bohn, H., Golatowski, F.: Service-Oriented Architectures for Embedded Systems Using Devices Profile for Web Services. In: 21st International Conference on Advanced Information Networking and Applications Workshops, AINAW 2007, Niagara Falls, Canada (2007)
8. Driscoll, D., Mensch, A.: Devices Profile for Web Services Version 1.1. OASIS (2009)
9. Simonis, I.: OGC Sensor Web Enablement Architecture. Open Geospatial Consortium, Inc. (2008)
10. Petak, W.J.: Emergency Management: A Challenge for Public Administration. Public Administration Review 45, 3–7 (1985)
11. Currion, P., de Silva, C., de Walle, B.: Open source software for disaster management. Commun. ACM 50, 61–65 (2007)
12. Doeweling, S., Probst, F., Ziegert, T., Manske, K.: Soknos - An Interactive Visual Emergency Management Framework. GeoSpatial Visual Analytics, pp. 251–262 (2009)
13. Zeeb, E., Bobek, A., Bohn, H., Prüter, S., Pohl, A., Krumm, H., Lück, I., Golatowski, F., Timmermann, D.: WS4D: SOA-Toolkits making embedded systems ready for Web Services. In: Open Source Software and Productlines 2007 (OSSPL 2007), Limerik, Ireland (2007)
14. Reichenbach, F.: Ressourcensparende Algorithmen zur exakten Lokalisierung in drahtlosen Sensornetzwerken. PhD thesis, University of Rostock, Rostock (2007)
15. Yang, Z., Liu, Y.: A Survey on Localization in Wireless Sensor networks. Hong Kong University (2005)
16. Steggles, P., Gschwind, S.: The Ubisense Smart Space Platform. A Ubisense White Paper. Dortmund (May 2005)

Hallway Monitoring: Distributed Data Processing with Wireless Sensor Networks

Tobias Baumgartner, Sándor P. Fekete, Tom Kamphans, Alexander Kröller, and Max Pagel

Braunschweig Institute of Technology, IBR, Algorithms Group, Germany
{t.baumgartner,s.fekete,t.kamphans,a.kroeller,m.pagel}@tu-bs.de

Abstract. We present a sensor network testbed that monitors a hallway. It consists of 120 load sensors and 29 passive infrared sensors (PIRs), connected to 30 wireless sensor nodes. There are also 29 LEDs and speakers installed, operating as actuators, and enabling a direct interaction between the testbed and passers-by. Beyond that, the network is heterogeneous, consisting of three different circuit boards—each with its specific responsibility. The design of the load sensors is of extremely low cost compared to industrial solutions and easily transferred to other settings. The network is used for in-network data processing algorithms, offering possibilities to develop, for instance, distributed target-tracking algorithms. Special features of our installation are highly correlated sensor data and the availability of miscellaneous sensor types.

Keywords: Sensor Networks, Testbeds, Data Processing, Target Tracking, Load Sensors.

1 Introduction

In the research field of wireless sensor networks, a tremendous amount of fundamental work over the past years has focused on protocol design and algorithm development. This has led to a high availability of common routing [3], time-synchronization [17], localization [4], and clustering [1] algorithms—often designed for general sensor network topologies. Similarly, many testbeds [9,15,18] were built during that time to run these algorithms on real sensor nodes. This became possible due to both dropping hardware costs and the maturing of operating systems running on the nodes, simplifying the development process. Due to the mainly common demands of the algorithms, most of the available testbeds were also held generic; the main focus was on the principal functionality of the algorithms and protocols, the aim being real-world communication behavior and implementations on tiny micro-controllers.

With the ongoing progress of algorithmic methods and system technology, it becomes possible as well as important to apply the previously designed basics to real application areas—thereby often adapting a generic solution to the specific needs of a single deployment. Such application areas for wireless sensor networks are quite dispersed. Deployments vary from monitoring environmental

P.J. Marron et al. (Eds.): REALWSN 2010, LNCS 6511, pp. 94–105, 2010.

areas such as volcanos or mountain sides, over personal area networks in medical applications, to home automation systems.

Building such real-world applications with actual sensor data processing is still a challenging task. First, the installation of specialized sensors often requires a significant amount of additional work. Second, such sensors may also cost much more than the nodes themselves—and thus are often not affordable for ordinary sensor network testbeds.

The design, development, and evaluation of higher-level algorithms in real deployments in which sensor nodes can share their local knowledge to obtain global goals requires appropriate sensor data. To carry out such tests, we developed a hallway monitoring system, consisting of 120 load sensors deployed beneath the hallway floor, and 29 passive infrared sensors (PIRs) for motion detection. The construction of the load sensors has already been demonstrated in [5]. The sensors are connected to nodes, which in turn can then exchange the measured values. The data is highly correlated, therefore serving as an ideal testbed for any algorithm performing data aggregation or in-network data analysis, such as distributed tracking algorithms.

The floor consists of square floor tiles with a side length of 60 cm each, which are installed on small metal columns. The setup is shown in Fig. 1.

We installed one load sensor on each of these columns. Therefore the corners of four floor tiles rest on each sensor, and vice versa each floor tile is monitored by four sensors. Every four load sensors are connected to a sensor node, which is also installed beneath the floor. Altogether, the setup consists of 120 load sensors, 29 PIR sensors, and 30 sensor nodes. The hallway has a width of 3 meters (corresponding to 5 tiles), and a length of 21.6 meters.

We designed the load sensors ourselves, with a surprisingly cheap construction. One load sensor costs about 25 Euros—as opposed to around 200 Euros for industrial manufactured load cells. The lower price comes with a loss of accuracy, but this loss can be compensated by sophisticated algorithms for sensor networks, where the nodes do in-network processing of the highly correlated data.

The rest of the paper is structured as follows. Section 2 describes similar constructions and related work. In Section 3, the hallway monitoring system is presented in detail. In Section 4, we report on how the sensor network can be accessed by the public. Section 5 describes first experimental results with the load sensors. We conclude the paper in Section 6.

2 Related Work

The development of a sensing floor has been proposed by other authors, but not in the context of a senor network, which is crucial for high-level methods and applications. Addlesee et al. [2] present a design with 3x3 tiles placed on load cells. Similarly, Orr and Abowd [13] designed the Smart Floor, also based on load cells. Neither of the authors considered a sensor network scenario.

While the above approaches make use of expensive load cells, Yiu and Singh [19] and Kaddoura et al. [10] presented designs based on force sensors. Like in

(a) The installation site.

(b) Floor tiles rest on columns.

Fig. 1. Hallway monitoring scenario

the other cases, there is no distributed data processing, and the system allowed only for presence detection, as opposed to more complex information such as the fine-grained resolution of a single step.

Mori et al. [11,12] present both a sensing room with pressure sensors on the floor and also the furniture, and a sensing floor, which they use to identify people via their gaits. The latter, gait recognition for people identification, has also been done by Qian et al. [14]. Again, there is no distributed in-network analysis by small devices like sensor nodes. In contrast to the previous descriptions, we present a both simple and highly affordable solution for hallway monitoring. In addition, our construction allows for the design of sophisticated algorithms running on tiny sensor nodes.

In general, the possibility of target tracking in indoor environments is especially interesting in the field of Ambient Assisted Living (AAL). Elderly, impaired, or disabled people are to be supported by technical solutions integrated in their homes. Gambi and Spinsante [7] present a localization and tracking system based on multiple cameras. In [8], Jin et al. analyze the performance of using accelerometers for AAL. However, having multiple cameras at home comes with a certain discomfort, as well as the need of constantly wearing sensors when at home. Our approach can work fully transparent for inhabitants, by offering the same features as above.

3 Hallway Construction

We have built a hallway monitoring system in our institute. To this end, we designed 120 load sensors, which were installed beneath the floor tiles in our hallway. A single load sensor and an exemplary section of the installation are shown in Fig. 2. There are also 29 PIR sensors on the walls to allow combining different kinds of sensor values in one distributed application. The sensors are connected to a total of 30 iSense [6] nodes, which can communicate over their radio. Finally, there are also actuators installed—29 light-emitting diodes (LEDs) and speakers to play sound samples—that are controlled by the sensor nodes, and thus enhance debugging possibilities of newly designed algorithms. A schematic diagram of the whole hallway construction with the interconnections of the several components is shown in Fig. 3. In the following, each part is described in detail.

3.1 Load Sensors and PIRs

We present a simple—and most notably low-cost—mechanism of building a single load sensor for our application. We use strain gauges, which are able to measure minimal strains in the objects to which they have been glued to. These strain gauges are supplied with a voltage of a few Volts, whereby they provide an output voltage of just a few millivolts. Whenever the attached material is strained or deformed, even by a few nanometers, the output current changes. Such sensors cost only a few Euros (around 10 Euros apiece in our case).

The strain gauges are attached to small steel plates with a size of approx. 10x4 cm. We used spring steel for the base construction. The advantage of spring steel is that it is flexible enough to be strained by the weight of a person, but also solid enough not to be permanently deformed. Installing the steel plates under the floor is surprisingly difficult. Strain gauges measure strains in different

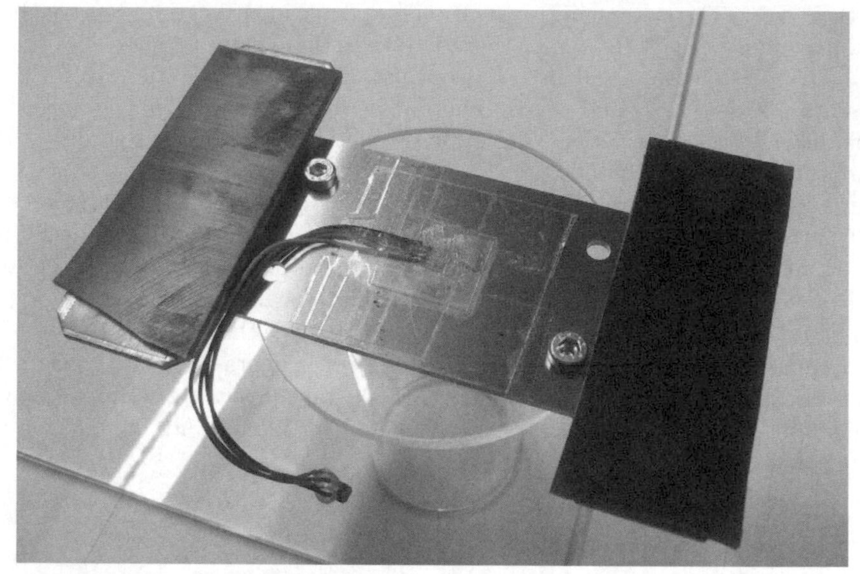

(a) A single load sensor.

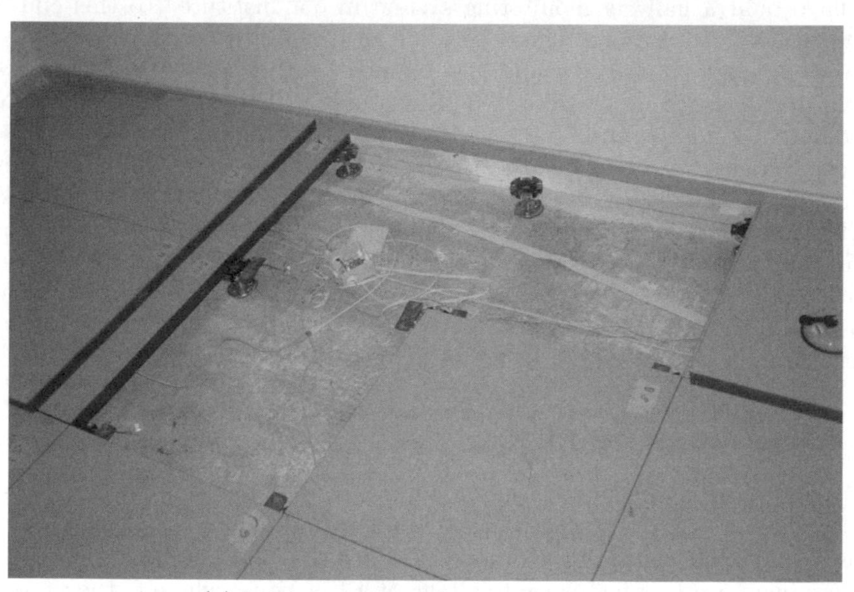

(b) Load sensors attached to iSense node.

Fig. 2. Load sensor installation

directions. If strain is applied from perpendicular directions, they annihilate each other, and the sensor does not measure any force. Hence, we enhanced the construction to deal with this issue. We use two additional steel plates, each with a spacer. The final construction is shown in Fig. 2(a). Like the strain gauges, the steel construction is surprisingly cheap. We paid around 15 Euros apiece.

Finally, the load sensors were installed in the hallway. The floor consists of square floor tiles with a side length of 60 cm each, which rest on small metal columns. We installed one load sensor on each of these columns—resulting in a total of 120 sensors beneath the floor. The sensor data is highly correlated, since the corners of four floor tiles rest on each sensor, and vice versa each floor tile is monitored by four sensors. The setup is shown in Fig. 2(b), where three floor tiles were removed to provide a view of the installation beneath the floor.

In addition to the load sensors beneath the floor, we installed 29 PIR sensors for motion detection on the walls. Each sensor is placed in a height of 2.5m, directed approx. 45° downwards. There are always two sensors face to face with each other, enabling the observation of a section of the hallway.

This facilitates identifying people by weight, but also by their motion when passing through the hallway. The different kinds of sensor values can then be combined to allow for the development of algorithms for heterogeneous sensor types.

3.2 Actuators

In contrast to the load sensors and PIRs, we have also added actuators to the hallway. There is a total of 29 lights and speakers installed on the walls. Each actuator consists of a so-called media board, which is an extra circuit consisting of an Atmega48, nine LEDs (three red, three green, three blue) attached to a cooling element, a speaker connected to the PWM output of the microcontroller, and a 4 GB microSD card for storing sound samples played through the speaker. Each media board is connected via UART to one sensor node beneath the floor, and can thus be used as an actuator for direct interaction with people passing the floor.

3.3 Sensor Nodes

Since the strain gauges have an output of just a few millivolts, they cannot be measured using ordinary ADCs. We use an additional amplifier circuit, to which up to six strain gauges can be attached. The circuit can power the sensors, and also read out and amplify the sensor output. It bears an Atmega48, which provides multiple ADC ports to read out the sensor values. The circuit has been designed to be used directly with our iSense sensor node platform [6], and communicates with the Atmega48 on the amplifier circuit via SPI. Even though it is iterated over up to six ADCs on the Atmel, and the data is additionally transmitted via SPI, we achieve a data transfer rate of 800 Hz per load sensor. This allows for highly fine-grained data-processing, and can lead to analyzing even single steps of passing people.

In addition to the connection to the load sensors, the iSense nodes are also wired to the PIR sensors and actuators on the wall. The whole setup is shown in Fig. 3. Each wireless sensor node is connected to four load sensors, one PIR, and one actuator unit—due to a diverging corridor one wall installation is missing, resulting in one iSense node without a PIR and LED/speaker unit connected.

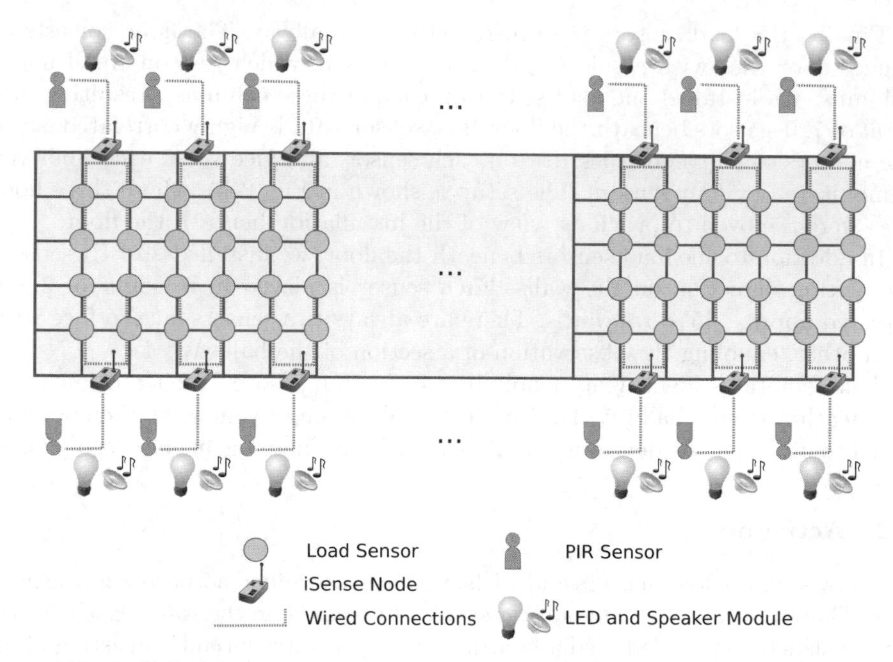

Fig. 3. Hallway construction with different kinds of sensors and actuators

The iSense nodes can then be used for the implementation of high-level data processing algorithms. For example, by exchanging actual data over the radio, the nodes can track people walking through the hallway.

For debugging purposes, the iSense nodes are connected via USB to a backbone of several PCs. The nodes are powered via this connection. In addition, they can be re-programmed, and debugging data can be collected continuously and reliably.

4 Software Access

There are two possibilities of accessing the sensor nodes in the hallway: First, there is an open API offered via web services. Second, we implemented a Java-based GUI for simple and fast algorithm development offering a central view on the network.

4.1 WISEBED API

The testbed was built in the context of the EU-project WISEBED [16], which aims at the interconnection of different sensor network testbeds spread over Europe. One goal of the project is to allow the connection of several testbeds and make them appear as only one testbed for a user. Moreover, we aim at allowing users to connect their own testbeds to one that is part of WISEBED. Therefore, all APIs that are needed to access a testbed and its sensor nodes are open to the public. Sensor nodes can be re-programmed, messages can be sent to the nodes, and debugging output can be collected. All APIs are based on web services for platform independence.

Since our testbed is part of the WISEBED project, our hallway monitoring system will be made available for the public—there is, of course, also a user management and reservation system offered.

4.2 CoCoS - Java-Based GUI

To facilitate the access to the sensor floor and enable users to develop their own software, we developed a simple to use Java API which gives access to the hallway. The so-called "Corridor Control System" (CoCoS) consists of a client-server solution which allows multiple clients to access the floor simultaneously. The server is embedded as a pluggable module in the WISEBED API and is able to fully control the floor.

CoCoS provides a real-time global view of the sensor floor, which can be easily accessed to program custom extensions, evaluate sensor data, or send commands to the sensor floor. Another feature is to write out sensor data traces, which can be played back later to run different algorithms on the same data, or work off-line without a connection to the hallway. The server does not provide a graphical user interface, but it is possible to connect a GUI-client to the server via TCP/IP that offers a graphical visualization of the current floor status, see Fig. 4. It is also possible to start an extension from the client to remotely control the corridor, which makes it possible to work with the testbed from anywhere.

Fig. 4. CoCoS, a Java-based GUI for accessing the hallway data

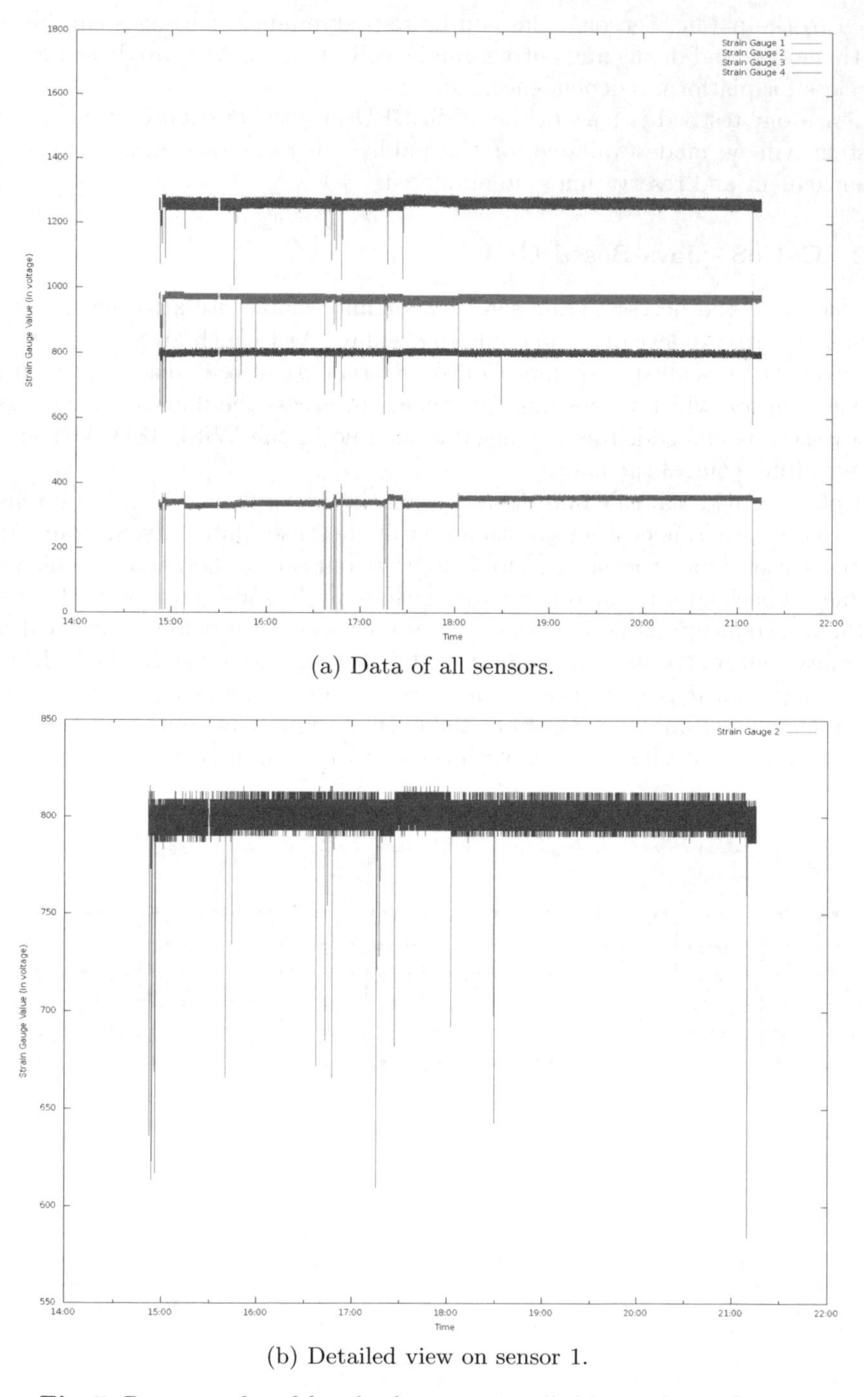

(a) Data of all sensors.

(b) Detailed view on sensor 1.

Fig. 5. Data samples of four load sensors installed beneath one floor tile

The advantage of CoCoS is that it offers a global view of the whole network and all sensor data. This simplifies the development process tremendously, since new ideas can be implemented and evaluated easily by a centralized algorithm written in Java, and then later translated to a distributed one working directly on the hallway nodes.

5 Experimental Study

We recorded data of four sensors that are installed beneath one floor tile to show the correlation of the values, and to have a look into the data of one sensor in detail. The data was collected with 8 Hz (we can also take samples with 800 Hz), since it was recorded over several hours. One data sample is the value read at the ADC of the Atmel connected to the iSense node, and hence the already amplified strain gauge value in voltage. When there is no load produced, the sensor value stays constant. Whenever there is load detected, the value drops by a certain amount. The results for the four sensors are shown in Fig. 5(a).

The zero value of the sensors differs significantly—from around 400 up to approx. 1300. This is due to the self-construction of the sensors, since the zero value depends on the force the strain gauge is glued on the steel plate. Analogously, the amplitude is different from sensor to sensor. Issues arising from differences in zero value or amplitude can be overcome using appropriate distributed algorithms. The important observation is the high correlation in the data, which can be seen in the synchronous amplitude changes of the four sensors.

Fig. 5(b) shows the trace of one sensor in detail. The data is the raw output from the strain gauge, and thus basic noise can be seen even when there is no force put on the sensor. However, one can clearly distinct an amplitude from the noise.

6 Conclusion and Future Work

We presented a hallway monitoring system based on load and PIR sensors, which are connected to wireless sensor nodes. The sensor data is highly correlated, and enables the design of sophisticated distributed algorithms for target tracking or gait recognition. The nodes can collaborate to substitute the merely imprecise data of the load sensors. The inaccuracy of the sensors is outweighed by the extremely low cost—about 25 Euros per sensor, in contrast to more than 200 Euros for industrial solutions.

In addition, we also added actuators to the testbed. We installed 29 lights and speakers on the walls to enable the possibility of interaction between the sensor network and passing people. Both lights and speakers can be controlled by the sensor nodes beneath the floor, and can thus be directly integrated in distributed applications.

The whole design also deals with heterogeneity. We have 30 iSense sensor nodes that can communicate wirelessly. Each node is connected to a circuit board equipped with an Atmel Atmega48, responsible for amplifying and receiving the load sensor data. This board is in turn wired to a circuit board at the walls, controlling the lights and speakers.

At this point, we finished the construction of the hallway. All sensors are installed, and the nodes are connected via USB to a backbone for reliable reprogramming and data collection. We have also evaluated the output of the load sensors. In the next step, we will develop algorithms for more challenging tasks, such as accurate target tracking, identification of the number of people in the hallway, or the study of different gaits when the available sample rate of 800 Hz per load sensor is considered.

Acknowledgement. This work has been partially supported by the European Union under contract number ICT-2008-224460 (WISEBED). We thank Marcus Brandenburger, Peter Degenkolbe, Henning Hasemann, Winfried Hellmann, Björn Henriks, Roland Hieber, Peter Hoffmann, Hella-Franziska Hoffmann, Daniel Houschka, Rolf Houschka, Andreas König, Christiane Schmidt, Nils Schweer, Stephan Sigg, and Christian Singer for their assistance in the construction of the hallway.

References

1. Abbasi, A.A., Younis, M.: A survey on clustering algorithms for wireless sensor networks. Computer Communications 30(14-15), 2826–2841 (2007)
2. Addlesee, M., Jones, A., Livesey, F., Samaria, F.: The ORL active floor [sensor system]. IEEE Personal Communications 4(5), 35–41 (1997)
3. Akkaya, K., Younis, M.: A survey on routing protocols for wireless sensor networks. Ad Hoc Networks 3(3), 325–349 (2005)
4. Akyildiz, I.F., Su, W., Sankarasubramaniam, Y., Cayirci, E.: Wireless sensor networks: a survey. Computer Networks 38, 393–422 (2002)
5. Baumgartner, T., Fekete, S.P., Kröller, A.: Hallway monitoring with sensor networks. In: Proceedings of the 7th ACM Conference on Embedded Network Sensor Systems, SenSys 2009, pp. 331–332. ACM, New York (2009)
6. Buschmann, C., Pfisterer, D.: iSense: A modular hardware and software platform for wireless sensor networks. Technical report, 6. Fachgespräch Drahtlose Sensornetze der GI/ITG-Fachgruppe Kommunikation und Verteilte Systeme (2007)
7. Gambi, E., Spinsante, S.: Multi-camera localization and tracking for ambient assisted living applications. In: AALIANCE Conference, Malaga, Spain (March 2010)
8. Jin, A., Yin, B., Morren, G., Duric, H., Aarts, R.: Performance evaluation of a tri-axial accelerometry-based respiration monitoring for ambient assisted living. In: Annual International Conference of the IEEE Engineering in Medicine and Biology Society, EMBC 2009, pp. 5677–5680 (September 2009)
9. Johnson, D., Stack, T., Fish, R., Flickinger, D.M., Stoller, L., Ricci, R., Lepreau, J.: Mobile emulab: A robotic wireless and sensor network testbed. In: INFOCOM. IEEE, Los Alamitos (2006)
10. Kaddoura, Y., King, J., Helal, A.S.: Cost-precision tradeoffs in unencumbered floor-based indoor location tracking. In: 3rd International Conference on Smart Homes and Health Telematics, From smart homes to smart care: ICOST (2005)

11. Mori, T., Sato, T., Asaki, K., Yoshimoto, Y., Kishimoto, Y.: One-room-type sensing system for recognition and accumulation of human behavior. In: IEEE/RSJ International Conference on Intelligent Robots and Systems (2000)
12. Mori, T., Suemaou, Y., Noguchi, H., Sato, T.: Multiple people tracking by integrating distributed floor pressure sensors and WID system. In: IEEE International Conierence on Systems, Man and Cybernetics (2004)
13. Orr, R.J., Abowd, G.D.: The smart floor: a mechanism for natural user identification and tracking. In: Extended Abstracts on Human Factors in Computing Systems, CHI 2000, pp. 275–276. ACM, New York (2000)
14. Qian, G., Zhang, J., Kidané, A.: People identification using gait via floor pressure sensing and analysis. In: Roggen, D., Lombriser, C., Tröster, G., Kortuem, G., Havinga, P. (eds.) EuroSSC 2008. LNCS, vol. 5279, pp. 83–98. Springer, Heidelberg (2008)
15. Raychaudhuri, D., Ott, M., Secker, I.: Orbit radio grid tested for evaluation of next-generation wireless network protocols. In: Proceedings of the First International Conference on Testbeds and Research Infrastructures for the DEvelopment of NeTworks and COMmunities, TRIDENTCOM 2005, Washington, DC, USA, pp. 308–309. IEEE Computer Society Press, Los Alamitos (2005)
16. Seventh Framework Programme FP7 - Information and Communication Technologies. Wireless Sensor Networks Testbed Project (WISEBED), ongoing project since (June 2008), `http://www.wisebed.eu`
17. Sundararaman, B., Buy, U., Kshemkalyani, A.D.: Clock synchronization for wireless sensor networks: a survey. Ad Hoc Networks 3(3), 281–323 (2005)
18. Werner-Allen, G., Swieskowski, P., Welsh, M.: Motelab: a wireless sensor network testbed. In: Proceedings of the 4th International Symposium on Information Processing in Sensor Networks, IPSN 2005, USA, IEEE Computer Society Press, Los Alamitos (2005)
19. Yiu, C., Singh, S.: Tracking people in indoor environments. In: Okadome, T., Yamazaki, T., Makhtari, M. (eds.) ICOST. LNCS, vol. 4541, pp. 44–53. Springer, Heidelberg (2007)

senSebuddy: A Buddy to Your Wireless Sensor Network

Adi Mallikarjuna Reddy V, Kumar Padmanabh, and Sanjoy Paul

SETLabs, Infosys Technologies Ltd., Bangalore, India, 560 100
{adi_vanteddu,kumar_padmanabh,sanjoy_paul}@infosys.com

Abstract. The sensor data are used by end users using various IT applications. The typical application for monitoring, querying and controlling the deployed Wireless Sensor Network is complex in nature from application development point of view, owing to various resource limitations. Post-deployment nuances like firmware update, addition of new nodes or replacement of others are not so easy task, either. In recent past, efforts have been made to minimize the complexity of end user through desktop and web-based application. However, so far, instant messaging has not been tried for communication between an end user and physical mote, where an individual sensor node (or group of them) appearing as a buddy in the instant messaging contact list and one can talk to it. In this paper we are describing a system that we developed recently with the name *senSebuddy*, based on instant messaging technique to monitor, query and control the deployed WSN applications. We also describe the functionality of the prototype implementation of the system with Smack XMPP API and Openfire XMPP server [7]. We have carried out series of experiments to prove that one can chat with sensor mote like a normal human being, and mote can be programmed using gtalk without experiencing any delay.

1 Introduction

The wireless sensor network (WSN) has been graduated from the research labs and is being used in daily life now. The individual sensors are connected to the outside world via a base station. The data generated by the motes are ultimately used by end users. The application software are either available in base station itself or in the server connected to it depending upon the complexity of application and footprint of final code. The components of the applications can be broadly classified into two categories: the monitoring applications and the control applications. Mostly these applications are either standalone for one user or they are available as web applications. Web applications are used to access the same application remotely by one or more users. These applications can be accessed either through the conventional network (LAN or internet) or through mobile applications. Visualization part of the application is more bulky in nature. Some of the examples of such applications are Mote-view [1], SpyGlass NOSY [2], and SESAME [3]. These user applications have following limitations:

P.J. Marron et al. (Eds.): REALWSN 2010, LNCS 6511, pp. 106–112, 2010.

1. The control application of WSN is complex. Changing the operating parameter and to update the firmware and other communication stack inside the mote requires special skill set.
2. The access of complex web applications through mobile phone is not always feasible.
3. All base station cannot be connected to mobile phone. Thus complex application cannot run on mobile application.

For remote operation of the sensor network, researchers have tried only web application method.In this paper, a method for enabling a WSN application to converse with an instant messaging client is presented. Such method may help in establishing communication between the sensor nodes and a user, to either get live updates of data monitored by the sensor nodes or for controlling the functionality of sensor nodes. This method may help in querying with the sensor nodes to ascertain the status of the sensor nodes.

Rest of the paper is organized in the following way. Section 2 presents Instant Messaging Technology. The senSebuddy system is detailed in section 3 with its architecture and constituting subcomponents. Section 4 presents implementation details of senSebuddy with experimental results. Conclusion along with future work is presented in section 5.

2 Instant Messaging

Though Instant Messaging (IM) has come into existence in early 1990s, it had only chat rooms on web servers where group of people could interact with each other privately, until, ICQ in 1996 has come up with public instant messaging server that anyone could use. In 1997 with AOL entry into this market gave its users the true power of instant messaging. ICQ has been the most successful and it is the basis for most of the modern instant messaging utilities.

Most of the IM service providers use proprietary protocols for communication. Interoperability is one of the most promising features of the IM services. Gtalk allows communicating with AOL users, while Yahoo messenger allows communicating with MSN messenger users. Instant messaging can be broadly classified into two categories according to the use: Public Instant Messaging and Enterprise Instant Messaging. Unlike public instant messaging where the service provided is free of cost and open to all, enterprise instant messaging is meant for an organization where employees can chat, transfer files, and have voice-chat to collaborate on a project. Some of the enterprise IM service platforms are: IBM Lotus same time, item Oracle Beehive and Microsoft Office Communications Live Server. These platforms provide server software and client software installed on every user PC. Microsoft integrated its IM server capabilities into Microsoft exchange server to leverage the users information such as credentials etc.

2.1 Instant Messaging Bot

IM bot is a program that uses Instant Messaging as an application interface. IM users can add IM bot to their buddy list the same way they add friends, relatives, family and co-workers. These bots can be used to provide some lookup information by connecting to database or on regular basis they can update users with information like stock quotes, weather reports and any other relevant information. Most of the services are being "IMified" now.

2.2 Instant Messaging Bot vs. Web Based Applications

Traditional web applications have no means to convey the end user whether the web page displayed is changed or new content is added to it. Because of this, user may have to visit to see if the page has been changed wasting unnecessary network bandwidth. An IM bot, on the other hand has the power to notify the end user of changes to the content.

A web server has no knowledge of the users who are currently accessing a web page and there is no way to contact the user back. An IM bot, on the other hand knows who are currently accessing the application and can send the information instantly as it is updated. This would save bandwidth because the unnecessary requests are eliminated.

3 *senSebuddy*: The Proposed IM Messenger for Sensor Nodes

senSebuddy works along the lines of IM bot principles. It is more than just a bot in terms of its functionality. System architecture of senSebuddy is depicted in Fig. 1. This figure shows how different components are interconnected to each other in the system. Deployed wireless sensor network is interfaced with the senSebuddy server consisting of different software components which in turn connected to the IM Server. On the other side, users are connected to the IM server.

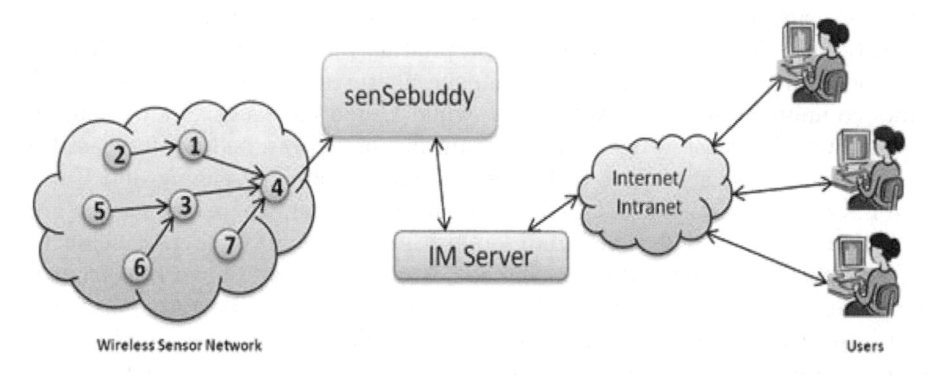

Fig. 1. senSebuddy Architecture

3.1 senSebuddy System

senSebuddy consists of three different software components. First one is senSebuddy middleware, second one is senSebuddy client and the last one is senSebuddy administration.

senSebuddy Middleware. senSebuddy Middleware is responsible for extracting the sensor data from the packets coming out of the WSN. Its architecture is depicted in Fig. 2. Once a packet is arrived at either serial/USB/Ethernet adaptor, the packet listener queues it for further processing. Filtering layers extracts the packet from the queue and decides whether it is a data packet or network related packet. If it is a control packet, then the network information is extracted out otherwise the packet is passed onto interpretation layer to decide upon what kind of sensor data it is having. Once the sensors information is deciphered, engineering conversion formulae are then applied on the raw data to extract the sensor readings. This is a one way communication from sensor network to the middleware. In some cases, user might want to send some data to the sensor network, in this case reverse communication layer plays a major role. Based on the user request it prepares a packet understood by the sensor network and sends it to the WSN through adaptor abstraction layer of middleware [8].

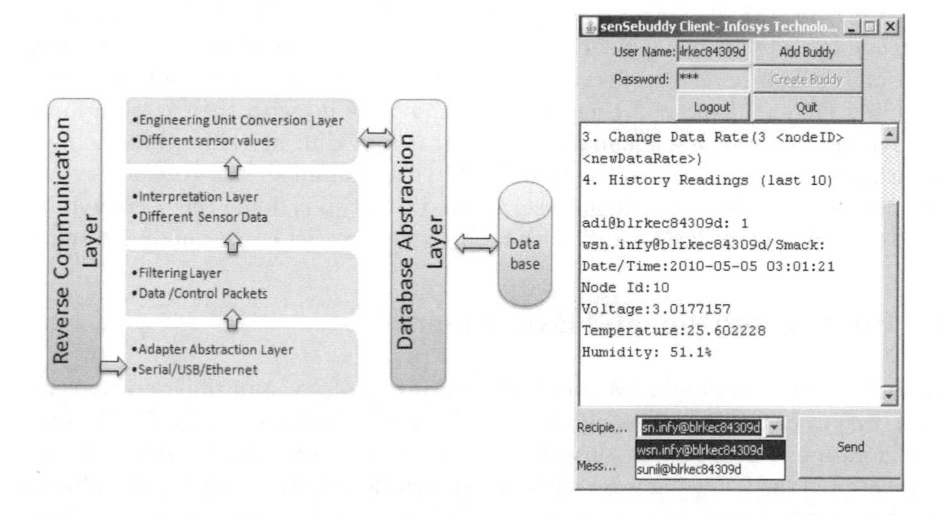

Fig. 2. senSebuddy Middleware Architecture diagram and a typical senSebuddy Client user interface

senSebuddy Client. senSebuddy client interacts with IM server to send and receive messages to/from users. This intercepts instant messages addressed to the deployed WSN nodes, based on the request type the corresponding reply message will be sent back to the user. Once the request is received, senSebuddy intercepts the messages, depending on the request type, the corresponding packet is formed and is sent to the middleware for further transmitting packet to the network. Once the response for the request is generated at the middleware, it further processes the packet and generates meaningful information out of it. Subsequently, senSebuddy client transforms the results into format understood by IM clients of users and sends the results to the corresponding request.

senSebuddy Administrator. An administration module is used for monitoring the administrating activity of the users interacting with the WSN. The validation process of the administration module may be used by an owner of the WSN to provide an administrative access to the WSN. The owner (herein also referred as "authorized user") may need to provide the consent for adding the messaging buddy of WSN with its unique identifier name by other users in their IM. The authorized user may register number of other users; those are entitled for accessing the WSN network through their IM. The process of adding the messaging buddy may include enabling the messaging buddy made available for communication with the other users through their instant messaging client. Administration module also provides selective access restriction to other user for communication with the one or more sensor nodes/one or more functionalities of the WSN.

3.2 senSebuddy Operation

Once the network is deployed, a unique user name is assigned in the IM server (e.g. *sensebuddy@gmail.com*). Depending on the requirement, the identity can be given to a single sensor node or collectively to entire network. Users can add the corresponding unique identity in their buddy list as they do with their conventional buddy name. Once the senSebuddy administrator module identifies that the user is authorized, it will add the user to its buddy list. Administrator can also set access control rights on a particular user. Once the setup is done, the users will be able to interact with WSN buddy with their requests to get corresponding results.

4 Implementation and Experiments

Though there are different IM standards and protocols exist in the literature, we have taken Extensible Messaging and Presence Protocol (XMPP) [4] into consideration. The different software components of senSebuddy system are implemented in the following way: senSebuddy middleware and senSebuddy administrator components are implemented using Java programming language. senSebuddy client is implemented using smack Java API [5]. Smack API is based on XMPP standard and can be used to connect to any IM server that supports XMPP standard. The screen shot of the same is depicted in Fig. 2. The developed senSebuddy client can be used to connect to GTalk as it supports the standard XMPP protocol for authentication, presence, and messaging [6].

We have deployed 10 XBow MicaZ motes with MDA300 sensor board in the conference rooms at our office premises. These sensor nodes sense room temperature and humidity at periodical intervals and communicate the same to the base station deployed at one of our cubicles. The data is logged in to the MySQL database at the server. We have implemented different request types and are explained below. Current Data: Request sent to know the current sensor readings. Actuation Request : Request sent to control the operation of appliances being monitored, Node Configuration (Data Rate) and History data Request : To know the history of sensor readings. Out of the four mentioned request types, except

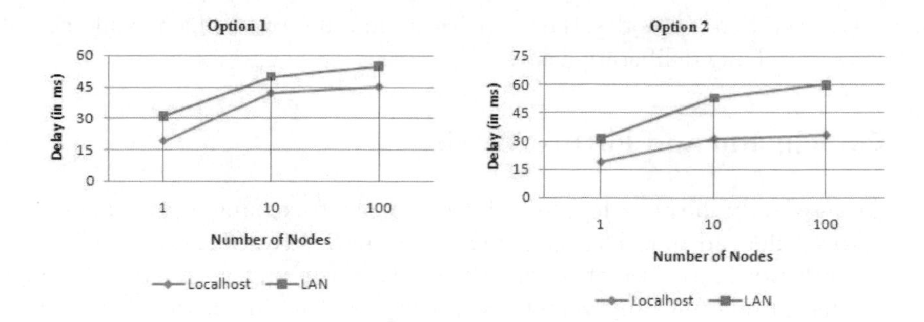

Fig. 3. Delay vs. No. Of Nodes for Option 1 and 2

the last one the other three requests have to be sent to the network for the results. The last request can be served with history of data logged in at the server.

We have used Openfire [7] XMPP Server as an IM Server. senSebuddy client developed using Smack API can be used to connect this IM server. We have given a unique id *wsn.infy@blrkec84309d.ad.infosys.com* to the deployed network and the same has been added by the fellow colleagues to their buddy list. We have not added any access restriction rules to this deployment and the users can access all the features of the network through their senSebuddy IM client.

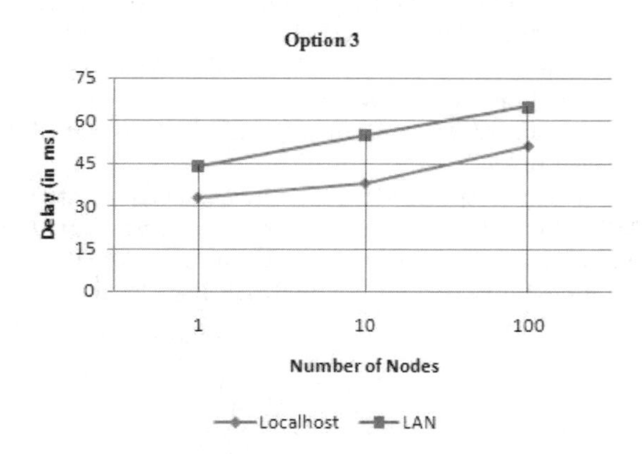

Fig. 4. Delay vs. No. Of Nodes for Option 3

We have conducted experiments with the above mentioned setup and the results are presented in the following figures. Option 1 refers to current data of a node in the network, option 2 refers to actuation request sent to a node in the network, and option 3 refers to altering the data rate of the node. Fig. 3 depicts the graph of option 1 and option 2 respectively with varying number of nodes in the network and the delay is plotted in the graph. Similarly fig 4 represents option 3. The 100 node experimentation is done by increasing the numbers of packets that a node sends in unit time. The above results indicate that with the increase in the number of nodes, the delay varies because of the latency involved in the network.

However, even with 100 nodes, the latencies are in milisecond which would not let the users to feel any deliberate delay.

5 Conclusion and Future Work

We have used only the basic feature of IM protocol i.e. exchanging messages. But, IM protocol offers advanced features such as file sharing, voice and video chat. We are currently working on remote updates to software running on the nodes through file sharing option of IM; this would ease the reprogramming process for end user also reduces the post-deployment maintainence activities. We are also planning to work on the voice feature of IM protocol as well so that users can interact with deployed WSN in very interactive manner.

References

1. Turon, M.: Mote-view: a sensor network monitoring and management tool. In: Proceedings of the 2nd IEEE workshop on Embedded Networked Sensors, EmNets 2005, pp. 11–17. IEEE Computer Society, Washington (2005)
2. Buschmann, C., Pfisterer, D., Fischer, S., Fekete, S.P., Kröller, A.: Spyglass: a wireless sensor network visualizer. SIGBED Rev. 2(1), 1–6 (2005)
3. S. Al-Omari, W. Shi, and C. J. Miller, Sesame: Sensor system accessing and monitoring environment, Wayne State University, Tech. Rep. MIST-TR-2004-018, 2004. [Online]. Available: http://www.cs.wayne.edu/ safwan/papers/sesame.pdf
4. XMPP, Xmpp foundation, http://www.xmpp.org
5. Smack, Smack api, http://www.igniterealtime.org/projects/smack/
6. Google Talk, Google talk, http://code.google.com/apis/talk/opencommunications.html
7. Openfire, Openfire, http://www.igniterealtime.org/projects/openfire/
8. Padmanabh, K., Malhotra, L., Reddy, A.M., Kumar, A.: MOJO: A Middleware that converts Sensor Nodes into Java Objects. In: IEEE ICCCN-2010 CON-WIRE, Zurich Switzerland (2010)

Evaluation of an Electronically Switched Directional Antenna for Real-World Low-Power Wireless Networks

Erik Öström, Luca Mottola, and Thiemo Voigt

Swedish Institute of Computer Science (SICS), Kista, Sweden

Abstract. We present the real-world evaluation of SPIDA, an electronically switched directional antenna. Compared to most existing work in the field, SPIDA is practical as well as inexpensive. We interface SPIDA with an off-the-shelf sensor node which provides us with a fully working real-world prototype. We assess the performance of our prototype by comparing the behavior of SPIDA against traditional omni-directional antennas. Our results demonstrate that the SPIDA prototype concentrates the radiated power only in given directions, thus enabling increased communication range at no additional energy cost. In addition, compared to the other antennas we consider, we observe more stable link performance and better correspondence between the link performance and common link quality estimators.

1 Introduction

The use of external antennas is a common design choice in many deployments of low-power wireless networks [13]. Indeed, an external antenna often features higher gains compared to the antennas found aboard mainstream devices, enabling increased reliability in communication at no additional energy cost. To implement such design, researchers and domain-experts have hitherto borrowed the required technology from WiFi networks [22, 10]. This holds both w.r.t. scenarios requiring omni-directional communication [22], and where the application at hand allows directional communication [10]. Although this implementation choice already enables improved performance, it is still sub-optimal in many respects, e.g., w.r.t. the significant size of the resulting devices, which complicates their installation. Unfortunately, as illustrated in Section 2, currently there are no *practical* solutions to address these issues, particularly in scenarios where some form of directional communication would be applicable.

To address this challenge, Nilsson designed SPIDA [11], an electronically switched directional antenna, shown in Figure 1. The SPIDA antenna is intended primarily for real-world low-power wireless networking, targeting scenarios that benefit from directional communication and sensor node localisation. We build a version of SPIDA that interfaces to a commercial sensor node—the popular TMote Sky platform [12]—and design and implement the software drivers necessary to dynamically control the direction of maximum gain. Section 3 describes the hardware/software integration of SPIDA with the sensor network platform.

We evaluate the performance of our SPIDA prototype in a real-world setting, as described in Section 4. We compare the SPIDA behavior against two omni-directional

P.J. Marron et al. (Eds.): REALWSN 2010, LNCS 6511, pp. 113–125, 2010.

Fig. 1. SPIDA prototype, connected to a TMote Sky node

antennas: an on-board micro-strip antenna and an external whip antenna for WiFi networks. We study the packet delivery rate and link quality using various network layouts, to assess communication ranges and directionality. To assess the dynamic abilities of SPIDA, we also run experiments by changing at run-time the direction of maximum gain. The results demonstrate that our SPIDA prototype behaves according to the intended design, and provides significant improvements in all metrics compared to the other antennas we consider.

The availability of a practical, inexpensive solution for dynamically controllable directional communication in low-power wireless networks raises interesting research questions and opens up a wealth of opportunities. We elaborate on this in Section 5, pointing to the network-level mechanisms that may leverage such antenna technology, and illustrating the expected performance gains.

We end the paper in Section 6 with brief concluding remarks.

2 Related Work

Nilsson identifies three candidate classes of directional antennas for low-power networks [11]: the adcock-pair antenna, the pseudo-doppler antenna, and the electronically switched parasitic element antenna. As described in Section 3, the SPIDA is an example of the latter class. At present, we could not find descriptions of other prototypes in any of these classes in the literature, let apart real-world experimental studies like ours.

The work closest to ours is that by Giorgietti et al. [8], who describe a prototype of four-beam patch antenna integrated with TMote Sky nodes, and related real-world experimental results. The direction of maximum gain is software-controlled, as in our SPIDA prototype. The size of the antenna, however, is much bigger than SPIDA. Giorgetti et al. leverage the experimental data to define analytical models for simulations. A similar activity using SPIDA is underway.

As already mentioned, antennas with fixed directions of maximum gain are employed in real-world applications [22, 10], but also as deployment tools. For instance, Saukh et al. [14] use "cantennas"—simple cylinder-shaped directional antennas—for node localisation and selective communication to a group of nodes.

Despite the lack of real-world prototypes of dynamically controllable directional antennas, the benefits they provide motivated research efforts at both MAC and routing layer [4, 5, 7, 15, 21], in low-power as well as mobile wireless networks. Most times, these leverage simulations or analytical studies based on abstract models of dynamically controllable directional antennas. Therefore, their behavior tends to be fairly idealized.

Advocating a top-down approach, some works provide guidelines for the design of dynamically controllable directional antennas based on the requirements imposed by higher-layer protocols [23, 19]. On the contrary, our research activity around the SPIDA antenna leverages a bottom-up approach, starting from a practical real-world antenna prototype, and then aiming at designing networking mechanisms leveraging its features, as discussed in Section 5.

3 Hardware/Software Design

In this section we describe the SPIDA hardware and the related control software.

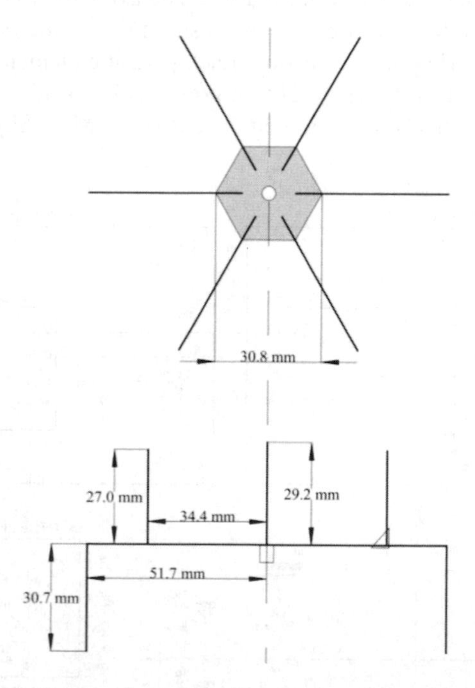

Fig. 2. SPIDA schematics without control electronics [11]

3.1 Hardware

The SPIDA antenna, developed at SICS by Nilsson [11], operates in the 2.4 GHz ISM band. SPIDA is a *switched parasitic element antenna* [18], i.e., it consists of a central active element surrounded by "parasitic" elements, as shown in Figure 2. The former is a conventional quarter-wavelength whip antenna. The parasitic elements can be switched between ground and isolation. When grounded, they work as reflectors of radiated power, and when isolated they work as directors of radiated power. The SPIDA is equipped with six parasitic elements, yielding six possible "switches" to control the direction of transmission.

A distinguishing feature is the SPIDA's smoothly varying radiation pattern. The antenna gain is designed to vary as an offset circle from approximately 7 dB to -4 dB in the horizontal plane, with the highest gain in the direction of the isolated parasitic elements. Although one may desire more selective transmission patterns, this choice simplifies the construction and use of the device, as we discuss in Section 5. In principle, such antenna behavior is obtained without any significant side lobes even when using simplistic on-off control [11]. The antenna is straightforward to manufacture, and its most expensive part is the SMA connector costing about 5 ECU in single quantities.

The circuitry to control the parasitic elements aims at reducing interference and suppressing noise from the sensor node digital circuitry. The schematics to control an individual parasitic element is shown in Figure 3. The available I/O lines on the TMote Sky are used to control the parasitic elements, using two LC filters for each I/O line to prevent noise from entering the RF section. Each parasitic element is controlled by an ADG902 SPST RF solid state switch. The control circuit is soldered onto a strip-board with an attached 10-pin IDC connector that fits onto the TMote Sky expansion pins.

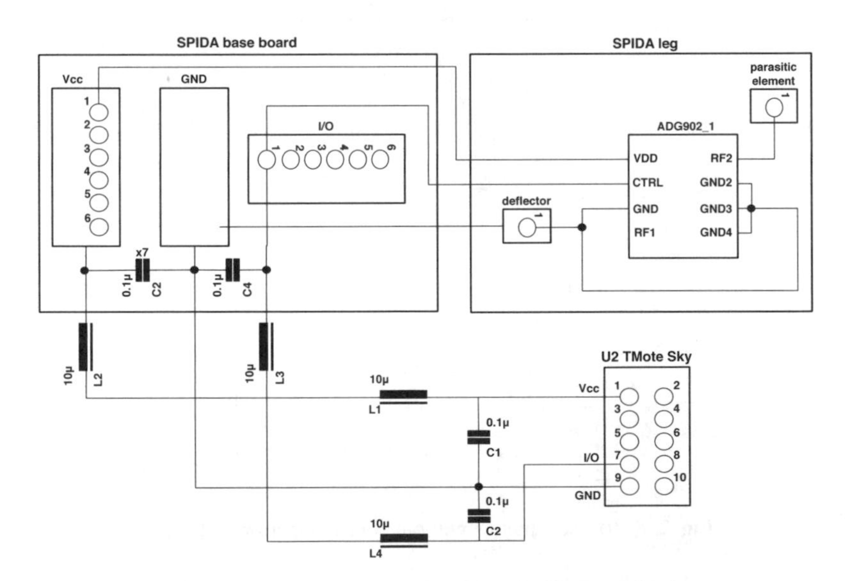

Fig. 3. SPIDA control electronics for a single parasitic element

Function	Input	Description
spida_init()	N/A	Initialize the driver.
spida_activate(int)	1-6	Isolate one of the six individual parasitic elements.
spida_deactivate(int)	1-6	Ground one of the six individual parasitic elements.
spida_configure(int)	0-6	Configure all parasitic elements at once to set a specific direction of maximum gain. (0 causes the SPIDA to behave as an omni-directional antenna).

Fig. 4. SPIDA driver API

Fig. 5. Test environment and antenna orientation on probe nodes

3.2 Software

We design and implement the software drivers necessary to control the six parasitic elements aboard the SPIDA, targeting the Contiki operating system [6]. The API provided to programmers is simple, as shown in Figure 4. The first function initializes the driver. The following two functions are used to isolate or ground specific parasitic elements on the SPIDA, enabling individual fine-grained control. Nevertheless, we expect the common use of the SPIDA to involve only one isolated element at a time, to direct the transmission in a specific direction. The last function in Figure 4 configures all parasitic elements at once to set a specific direction of maximum gain. Giving `0` as input makes the SPIDA isolate all parasitic elements, corresponding to omni-directional behavior. For instance, this may be useful for neighbor discovery.

4 Real-World Evaluation

We present the real-world evaluation we perform with our SPIDA prototype. Our objective is to investigate the SPIDA performance at the physical layer compared to the TMote Sky embedded microstrip antenna [20] and an external whip antenna for WiFi networks. The latter is connected to the node through a standard SMA connector and features a nominal gain of 2 dB.

4.1 General Setting

We deploy the nodes in an open grass field, shown in Figure 5. The location we choose has no interference coming from other networks working in the ISM band. We verify this condition by taking periodic noise floor measurements during the experiments, also with the TMote Sky's CC2420 radio chip. We install the nodes atop 1 m tall cardboard pillars to avoid signal reflections from the ground [3], and power them through the USB connector to factor out the influence of the battery discharge. All antennas we consider are oriented with the radiating element orthogonal to the ground, as shown at the bottom right of Figure 5. We carry out all experiments in comparable conditions of humidity and temperature. We check these conditions during the experiments by periodically querying the TMote Sky's integrated SHT11 sensor.

The various scenarios we investigate differ in the network layout, as described next. In every case, however, one node transmits using different antennas, while the others operate as passive probes, logging the received packets. The probes employ the external whip antenna shown in Figure 5. The SPIDA is always configured with only one parasitic element isolated: the configuration that yields the highest degree of directional transmission. For each experiment, the transmitter sends 1000 packets with an inter-packet interval of 500 ms. We use the lowest power setting, which enables easier logistics. The experiment code is implemented on top of the Contiki [6] operating system, and uses channel 26 for the transmissions.

As performance metrics, we consider averages over all probe nodes of the following figures: *i)* the *packet delivery rate* (*PDR*), defined as the average number of packets received at a probe over those sent by the transmitter, *ii)* the *received signal strength*

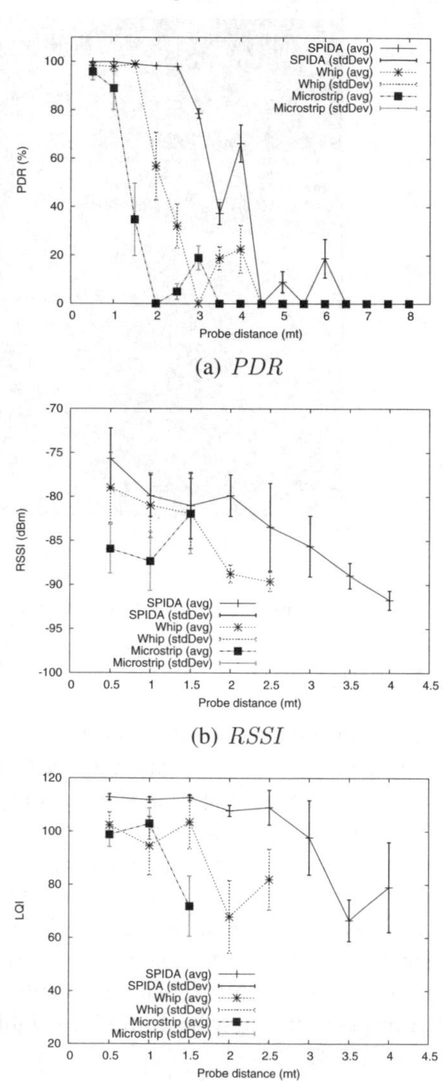

(c) *LQI*

Fig. 6. The SPIDA antenna extends the radio range and enjoys better correspondence between *LQI* and *PDR* compared to the other antennas

($RSSI$), and *iii)* the link quality indicator (LQI). We obtain the two latter for every *received* packet directly from the CC2420 radio chip. Because of this, the charts for $RSSI$ and LQI do not show regions where no packets were received. The results described next are averages over at least 5 repetitions of every experiment.

4.2 Network Layouts and Results

We describe next the specific network layout in every experiment and report on the corresponding results.

Range experiments. We compare the communication range of the SPIDA antenna against the other antennas we consider. To do so, we use only one probe node, placed at varying distances from the transmitter. In the first round of these experiments, the SPIDA has the isolated parasitic element pointing towards the probe.

Figure 6 illustrates the results. As shown in Figure 6(a), in the direction of maximum gain the SPIDA reaches much farther than the other two antennas. Using the SPIDA, the "connected" region [24] with PDR above 90% is about twice that of the whip antenna, and four times the case of the microstrip one. This is a key metric, as it indicates the portion of space characterized by reliable communication. The SPIDA also extends the "grey area" [24], characterized by highly varying performance and no predictable behavior. This is also an effect of the extended communication range.

The result above is reflected in the trends for $RSSI$ and LQI, shown in Figure 6(b) and 6(c). Moreover, within the connected region the SPIDA shows better correspondence between LQI and PDR than the other antennas. Thus, with comparable link performance in PDR, link quality estimators based on LQI [17] are likely to perform better with the SPIDA.

We also repeat the experiment with the isolated parasitic element of the SPIDA pointing in the direction opposite to the probe. Using this setting, the probe always receives less than 10 packets at 0.5 m from the transmitter, and then nothing beyond 1 m. This is a first evidence that the SPIDA does direct the transmitted power in a given direction. We investigate these aspects further in the following experiments.

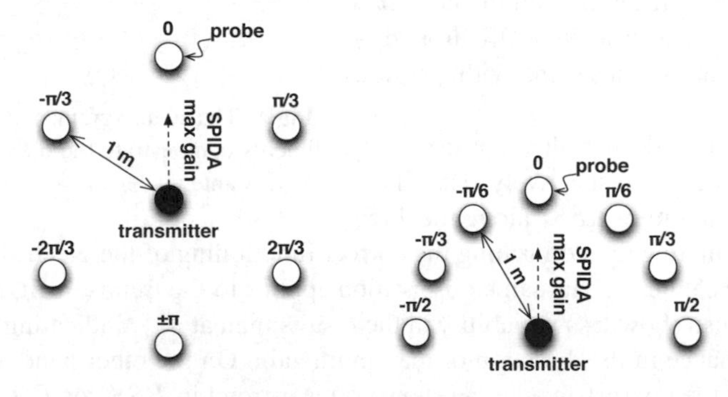

(a) Coarse-grained experiments. (b) Fine-grained experiments.

Fig. 7. Network layout for directional experiments

Coarse-grained directional experiments. We aim at a first, coarse grained characterization of the spatial characteristics of SPIDA transmissions compared to the other two antennas. To this end, we place the transmitter in the center of a circle of six probe nodes, as shown in Figure 7(a). Based on the results of the range experiments, we place the probes at 1 m from the transmitter, corresponding to the connected region for all antennas. We place the probes with the TMote Sky's USB connector pointing towards the transmitter. When using the SPIDA, every probe is aligned with a parasitic element.

We show the results in Figure 8. As depicted in Figure 8(a), the SPIDA achieves about 100% PDR only along the direction of maximum gain, corresponding to the isolated parasitic element. We also observe that the transmission pattern forms a lobe large enough to cover the probes at $\pm\frac{\pi}{3}$ as well, which still receive a significant number of packets. Nevertheless, the probes at $\pm\frac{2\pi}{3}$ and $\pm\pi$ receive no packets at all. This behavior largely corresponds to the simulation results reported earlier [11]. Thus, despite its simplicity, the electronics we built have very little influence on the antenna performance. As expected, the whip antenna shows an almost perfect omni-directional behavior. On the other hand, the microstrip antenna suffers from the co-location with the node base board, showing a drop in PDR around $\frac{\pi}{3}$. Such behavior is consistent with previous findings [20].

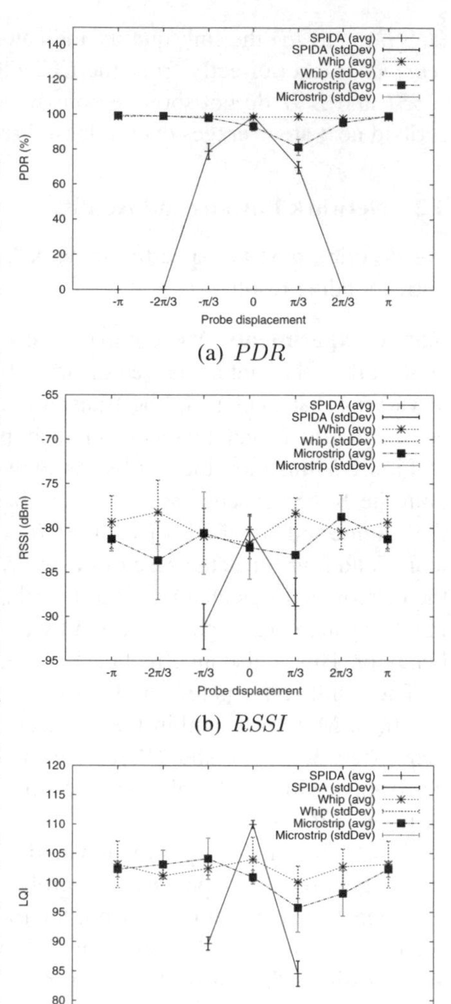

(a) PDR

(b) $RSSI$

(c) LQI

Fig. 8. The coarse-grained directional experiments demonstrate the directionality of the SPIDA antenna

Figure 8(b) and 8(c) illustrate the trends in $RSSI$ and LQI, respectively. The SPIDA shows a maximum in $RSSI$ along the direction of maximum gain, confirming the correct functioning of the electronics to control the parasitic elements. The same observation applies to the trends in LQI. Both points of maxima also show less variability in the results than at $\pm\frac{\pi}{3}$, indicating a more stable link performance in the direction of maximum gain. On the other hand, both the whip antenna and the microstrip antenna show no clear trend in $RSSI$ or LQI. When using omni-directional antennas, these metrics are known not to show a clear correspondence with PDR in most cases [16].

Fine-grained directional experiments. We investigate the transmission pattern of the SPIDA antenna at a finer grain around the direction of maximum gain. We deploy seven probes in a half-circle configuration, as in Figure 7(b). The other parameters are as in the previous coarse-grained experiments.

The results we obtain this time are shown in Figure 9. Figure 9(a) demonstrates the smoothly varying radiation pattern of the SPIDA. The PDR gradually decreases between 0 degrees—which is aligned with the isolated parasitic element—and $\pm\frac{\pi}{3}$, until it drops to zero at $\pm\frac{\pi}{2}$. Again the whip antenna behaves in an omni-directional manner, whereas the microstrip shows a larger drop around $\frac{\pi}{3}$, due to the higher spatial resolution of these experiments.

The trends in $RSSI$ and LQI, shown in Figure 9(b) and 9(c), confirm our observations. With the SPIDA, the decrease in both metrics is gradual around the direction of maximum gain, and the variability is reduced along this direction compared to both the other two antennas and the other directions with the SPIDA.

Dynamic experiments. We also test the SPIDA's ability to change the direction of maximum gain at run-time. We use again the network layout in Figure 7(a). However, this time we program the transmitter to switch the isolated parasitic element after *every* packet, moving the direction of maximum gain clockwise in the horizontal plane. We repeat this experiment 10 times.

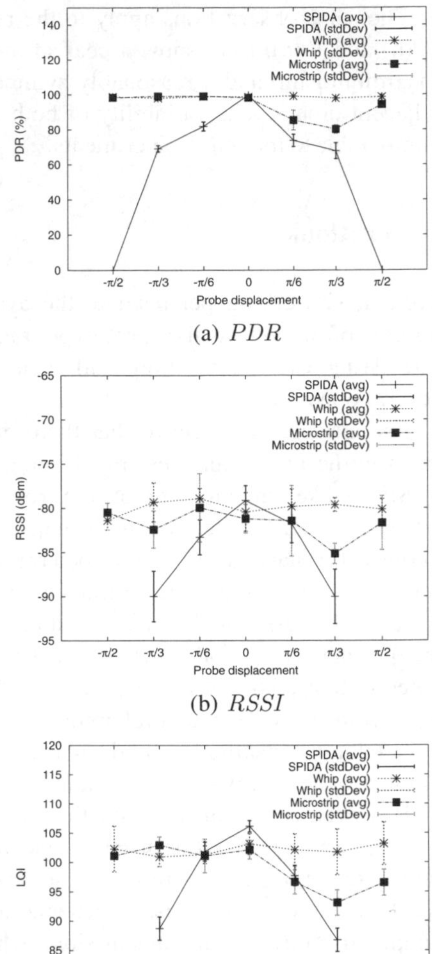

(a) PDR

(b) $RSSI$

(c) LQI

Fig. 9. The fine-grained directional experiments again demonstrate SPIDA's directionality w.r.t. all metrics

Figure 10 illustrates the trends in the metrics we consider as a function of a given probe, against the current direction of maximum gain. All results are remarkably consistent no matter which probe we examine. For instance, Figure 10(a) shows that all probes observe the same behavior in PDR as the direction of maximum gain changes, with the only difference of a variable offset due to a probe's relative displacement. It also appears that the SPIDA slightly favors the PDR at the probe to the left of the direction of maximum gain. This behavior is presumably due to some little imperfections in the construction process, which can be easily rectified.

The same observations apply to the results in $RSSI$ and LQI, depicted in Figure 10(b) and 10(c). Both show a peak at the probe aligned with the current direction of maximum gain, and a reasonably symmetric decrease of the same metric at the two adjacent probes. The variability of both $RSSI$ and LQI (not shown in the charts) is comparable to the other experiments.

5 Outlook

From a networking perspective, the availability of a SPIDA-like prototype raises interesting research questions and opens up several opportunities.

For instance, we believe that there may be significant advantages by leveraging a SPIDA-like antenna are at the routing layer. Consider the classical multi-hop, convergecast scenario using tree-shaped routing topologies. By using directed transmissions towards the parent node, one may diminish the probability of collisions due to simultaneous transmissions along parallel paths. This would provide greater reliability and reduce energy consumption by decreasing the number of necessary retransmissions.

However, achieving this functionality is not necessarily trivial. For instance, one may devise directionality-aware parent selection mechanisms, or re-use existing schemes and simply use directional transmissions when sending to the parent. In the latter case, the increase in communication range, which we also observed with the SPIDA in Section 4, may allow transmissions to reach non-parent nodes that are however closer to the sink. Significant trade-offs are involved in devising similar functionality, e.g., complexity vs. communication overhead, which deserve careful investigation.

Another example is related to the use of dynamically controllable directional antennas in TDMA-like MAC protocols. Doing so may enable spatial diversity in addition to time diversity. In this context, the few existing solutions tend to be very complex [21]. However, the SPIDA's radiation

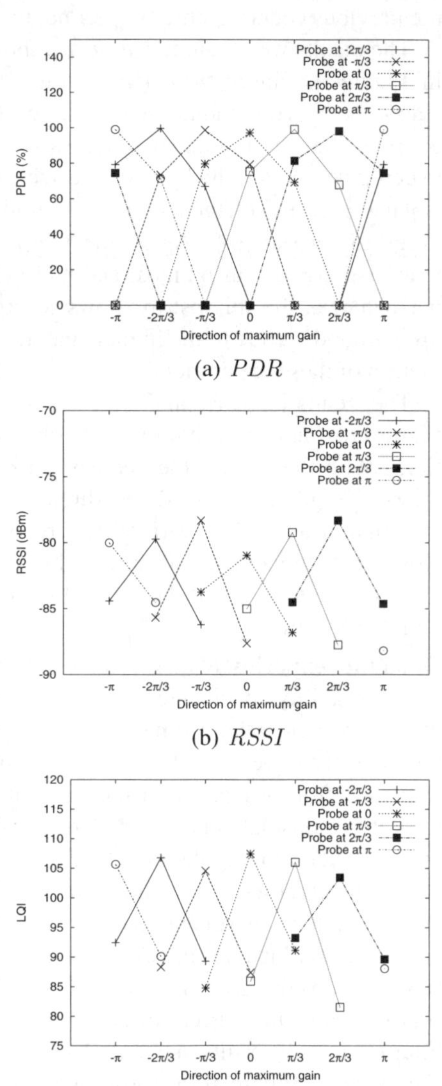

(a) PDR

(b) $RSSI$

(c) LQI

Fig. 10. The dynamic experiments demonstrate that the trends in PDR, $RSSI$, and LQI follow the changes in the direction of maximum gain

pattern, characterized by a simple offset circle, may greatly simplify the problem at the cost of slightly increased contention on the wireless medium. Here again, the trade-off between the degree of directional communication and the simplifications in the MAC operation shall be analyzed thoroughly.

Even staple networking mechanisms such as neighbor discovery may benefit form the use of dynamically controllable directional antennas. How to leverage this functionality, however, is an open question. If the antenna also provides omni-directional behavior, as in the case of SPIDA, one may re-use existing mechanisms. However, when the antenna turns to directional mode, the increased transmission range may reach nodes that were previously not recognized as neighbors. This would impact the operation of MAC protocols, as topology information would suddenly become inconsistent. Topology control schemes [9] may decrease the transmission power to maintain the same neighboring relations when the antenna is operating in directional mode. However, this would partly defeat the increased reliability obtained with directional transmissions.

On the other hand, one may use directional mode for neighbor discovery as well, rapidly sweeping all possible directions. However, by doing so, the link quality to different neighbors would be sampled at slightly different times, which might affect the operation of higher-level mechanisms, especially multi-hop routing protocols [1]. Most existing works in this area assume a priori knowledge on node positions. Even though directional antennas like the SPIDA are used for localization based on angle-of-arrival information [2], we do need much better integration of these functionality.

6 Conclusion

In this paper we reported on real-world experiments with SPIDA, an electronically switched directional antenna for low-power wireless networks. We showed that SPIDA concentrates the radiated power only in given directions. Based on a comparison with the on-board micro-strip antenna of the TMote Sky node and an external whip antenna, we observed increased communication range, improved link stability, and better correspondence between link performance and common link quality estimators. As we illustrated, this opens up several opportunities for improved network-level mechanisms that leverage the characteristics of SPIDA-like antennas.

Acknowledgements. We thank Martin Nilsson (SICS) who designed the SPIDA antenna and advised us on interfacing SPIDA to a TMote Sky sensor node. This work was supported by VINNOVA, the Uppsala VINN Excellence Center for Wireless Sensor Networks WISENET, also partly funded by VINNOVA, and CONET, the Cooperating Objects Network of Excellence, under EU-FP7 contract number FP7-2007-2-224053.

References

1. Al-Karaki, J., Kamal, A.E.: Routing techiniques in wireless sensor networks: A survey. IEEE Wireless Communications 11(6) (2004)
2. Amundson, I., Sallai, J., Koutsoukos, X., Ledeczi, A.: Radio interferometric angle of arrival estimation. In: Silva, J.S., Krishnamachari, B., Boavida, F. (eds.) EWSN 2010. LNCS, vol. 5970, pp. 1–16. Springer, Heidelberg (2010)

3. Anastasi, G., Borgia, E., Conti, M., Gregori, E., Passarella, A.: Understanding the real behavior of Mote and 802.11 ad hoc networks: An experimental approach. Elsevier Pervasive and Mobile Computing Journal 1(2), 237–256 (2005)
4. Bazan, O., Jaseemuddin, M.: On the design of opportunistic MAC protocols for multi-hop wireless networks with beamforming antennas. IEEE Transactions on Mobile Computing 99 (2010) (prePrints)
5. Cho, J., Lee, J., Kwon, T., Choi, Y.: Directional antenna at sink (DAaS) to prolong network lifetime in wireless sensor networks. In: Proc. of the European Wireless Conf. - Enabling Technologies for Wireless Multimedia Communications (2006)
6. Dunkels, A., Grönvall, B., Voigt, T.: Contiki - A lightweight and flexible operating system for tiny networked sensors. In: Proc. of 1st Wkshp. on Embedded Networked Sensors (2004)
7. Dunlop, J., Cortes, J.: Co-design of efficient contention mac with directional antennas in wireless sensor networks. In: Proc. of the International Wireless Communications and Mobile Computing Conference (2008)
8. Giorgetti, G., Cidronali, A., Gupta, S., Manes, G.: Exploiting low-cost directional antennas in 2.4 GHz IEEE 802.15.4 wireless sensor networks. In: Proc. of the European Conf. on Wireless Technologies (2007)
9. Hackmann, G., Chipara, O., Lu, C.: Robust topology control for indoor wireless sensor networks. In: Proc. of the 6th Int. Conf. on Embedded Networked Sensor Systems, SenSys (2008)
10. Kim, S., Pakzad, S., Culler, D., Demmel, J., Fenves, G., Glaser, S., Turon, M.: Health monitoring of civil infrastructures using wireless sensor networks. In: Proc. of the 6th Int. Conf. on Information Processing in Sensor Networks, IPSN (2007)
11. Nilsson, M.: Directional antennas for wireless sensor networks. In: Proc. of the 9th Scandinavian Workshop on Wireless Adhoc Networks, Adhoc (2009)
12. Polastre, J., Szewczyk, R., Culler, D.: Telos: Enabling ultra-low power wireless research. In: Proc. of the 5th Int. Conf. on Information Processing in Sensor Networks, IPSN (2005)
13. Raman, B., Chebrolu, K.: Censor networks: a critique of "sensor networks" from a systems perspective. SIGCOMM Comput. Commun. Rev. 38(3) (2008)
14. Saukh, O., Sauter, R., Meyer, J., Marrón, P.: Motefinder: A deployment tool for sensor networks. In: Proc. of the Workshop on Real-World Wireless Sensor Networks, REALWSN (2008)
15. Shihab, E., Cai, L., Pan, J.: A distributed, asynchronous directional-to-directional MAC protocol for wireless ad hoc networks. IEEE Trans. on Vehicular Technology 58(9) (2009)
16. Srinivasan, K., Dutta, P., Tavakoli, A., Levis, P.: An empirical study of low power wireless. ACM Transactions on Sensor Networks (2010) (to appear)
17. Srinivasan, K., Levis, P.: RSSI is under-appreciated. In: Proc. of the 3rd Int. Workshop on Embedded Networked Sensors, EmNets (2006)
18. Thiel, D., Smith, S.: Switched parasitic antennas for cellular communications. Artec House, London (2002)
19. Tian, Q., Bandyopadhyay, S., Coyle, E.J.: The effect of directional antennas on spatiotemporal sampling in clustered sensor networks. J. of Internet Technology (JIT) 8(1) (2007)
20. TMote Sky Datasheet, `www.snm.ethz.ch/pub/uploads/Projects/tmote_sky_datasheet.pdf`
21. Vilzmann, R., Bettstetter, C.: A survey on MAC protocols for ad hoc networks with directional antennas. In: Networks and Applications Towards a Ubiquitously Connected World, EUNICE 2005, vol. 196(1) (2006)

22. Werner-Allen, G., Lorincz, K., Johnson, J., Lees, J., Welsh, M.: Fidelity and yield in a volcano monitoring sensor network. In: Symp. on Operating Systems Design and Implementation, OSDI (2006)
23. Wu, Y., Zhang, L., Wu, Y., Niu, Z.: Interest dissemination with directional antennas for wireless sensor networks with mobile sinks. In: Proc. of the 6th Int. Conf. on Embedded Networked Sensor Systems, SenSys (2006)
24. Zhao, J., Govindan, R.: Understanding packet delivery performance in dense wireless sensor networks. In: Proc. of the 3rd Int. Conf. on Embedded Networked Sensor Systems, SenSys (2003)

Implementation and Evaluation of Combined Positioning and Communication

Paul Alcock, James Brown, and Utz Roedig

Lancaster University, UK
{p.alcock,j.brown,u.roedig}@lancaster.ac.uk

Abstract. A new generation of communication transceivers are able to support Time of Flight (TOF) distance measurements. Transceiver manufacturers envision that communication and positioning features are used separately and one at a time. However, we have demonstrated that such separation is unnecessary and that TOF measurements can be obtained during data communication. Thus, distance measurements required by positioning services can be collected both energy and bandwidth efficiently. In this paper we describe the modification of an existing low power wireless sensor network (WSN) medium access control (MAC) protocol called FrameComm to include collection of distance measurements. We describe the implementation of the modified FrameComm protocol (called FrameCommDM) on a Nanotron DK development kit comprising of an Atmel ATmega128 MCU and a Nanotron NA5TR1 transceiver. The performance of the FrameComm and FrameCommDM implementation on the Nanotron platform is evaluated and compared. It is shown that the collection of distance measurements has no significant impact on communication performance. Furthermore, the difference in energy consumption used to perform the additional ranging tasks of FrameCommDM is examined and quantified.

1 Introduction

Many positioning systems have been developed which use the existing communication transceiver of a sensor node. Positioning systems relying on conventional low-power communication transceivers typically make use of either the received signal strength (RSS) or the measured time-of-flight (TOF) of a signal as input for a positioning algorithm. Both methods can be used to determine the distance between transceivers and ultimately the position of all transceivers in relation to each other. These methods of distance measurements have been investigated at length and reports show that using current transceivers yield unreliable and inaccurate results. Patwari et al. [1] present an in-depth report of their findings, which show that multipath signals and shadowing obscure distance measurements.

The recent development of low-power, ultra wideband (UWB) transceivers for use in sensor nodes, overcomes the aforementioned ranging inaccuracies. The physical signal properties of UWB communication make it possible to accurately determine the time-of-arrival (TOA) of signals. By utilizing either clock synchronization or two-way-ranging it is therefore possible to accurately determine the time of flight (TOF) of the signal. Thus, the distance between communicating transceivers and node positions can

P.J. Marron et al. (Eds.): REALWSN 2010, LNCS 6511, pp. 126–137, 2010.

be determined. The IEEE 802.15.4a physical layer specification [2], standardized in 2007, defines the use of UWB transceivers for use in wireless personal area networks and the functionality of positioning. The Nanotron NA5TR1 [3] transceiver is an example of one such transceiver which adheres to this standard.

UWB transceiver manufacturers envision that communication and positioning features are used separately and one at a time. Either the transceiver is used to transfer data packets between sender and receiver *or* the transceiver is used to send ranging packets to determine the TOF between nodes. This leads to inefficient transceiver usage as excess packets might be generated unnecessarily. If an exchange of data packets is currently taking place between two nodes using a send and acknowledge scheme, the same packets can also be used to measure TOF between the nodes. Therefore the distance can be estimated using the existing data packets, alleviating the need to transmit specialized ranging packets.

We have shown in our previous, simulated work [10] how standard low power MAC protocols for WSNs can be modified to perform ranging while transmitting data. In particular, our previous work shows how the existing FrameComm MAC protocol [4] can be extended to support positioning tasks. In this paper we take our work one step further and describe a real world implementation of the modified FrameComm MAC protocol (called FrameCommDM) for the Nanotron NA5TR1 transceiver. This paper has the following specific contributions:

- Implementation Details: A detailed description of the FrameCommDM implementation for the Nanotron platform is given. We identify potential hardware changes to the Nanotron platform which would allow us to implement combined positioning and communication more efficiently.
- Prototype Evaluation: A comprehensive evaluation of the prototype system is given. In particular the impact of positioning on communication performance and system power consumption is quantified.

The prototype evaluation supports our previous results obtained via simulation (see [10]). Positioning features can be integrated into a low power WSN MAC protocol without significant impact on communication performance.

The next Section gives an overview on related work. Section 3 describes FrameComm and FrameCommDM. Section 4 details the implementation of FrameCommDM on the Nanotron platform. Section 5 outlines the evaluation setup and explains the obtained measurement results. The paper then concludes in Section 6 and describes proposed future work.

2 Related Work

There is a large body of work which has focused on exploiting UWB for either communication or positioning in wireless sensor networks. However, there is little research on how to tightly integrate both positioning and communication functions.

Correal et al. [5] present a method of positioning using UWB transceivers in which a packet sent by a node is followed by acknowledgements which can be used to derive round-trip times. This method provides a compelling proof-of-concept for our proposed system. However, the method discussed by Correal et al. differs from our method in that ranging is not formally integrated into the protocol, and there is no analysis of how their methods of positioning and communication functions affect one another.

Cheong and Oppermann [6] describe a positioning-enabled MAC protocol for UWB sensor networks. First, their solution differs from our work as data packets themselves are not used to support positioning; positioning and communication are handled completely separately by the MAC layer. Second, Cheong's work proposes a TDMA protocol, while the modified FrameComm protocol presented in this paper is a contention-based protocol.

The IEEE 802.15.4a physical layer specification [2], standardized in 2007, defines the use of UWB transceivers in wireless personal area networks. The standard defines positioning and communication as separate functions but does not discuss their integration. However, modern packet-based transceivers conforming to the 802.15.4a standard could potentially be used to support the MAC protocol defined in this paper.

The use of packetized radios requires a fresh approach of implementing asynchronous duty cycles in WSNs. Some schemes use the same concept of framelet trails as used by the FrameComm [4] MAC protocol used for the work presented in this paper. The current default energy saving protocol in TinyOS is based on the Low Power Listening component of BMAC [7], but employs message retransmission instead of a long preamble in order to accommodate packet-based radios. X-MAC [8] also uses framelets to establish rendezvous between sender and receiver but only retransmits the message header. The payload is sent only after one of the headers has been acknowledged by the destination. These and other existing framelet based MAC protocols can potentially be used in conjunction with UWB transceivers to integrate positioning and communication. Hence, the basic mechanisms described in this paper are not limited to the particular MAC protocol we have chosen (FrameComm).

3 FrameComm and FrameCommDM

This section describes the most important aspects of FrameComm and the ranging enabled variation FrameCommDM. A detailed description of FrameComm and FrameCommDM can be found in [4] and [10] respectively.

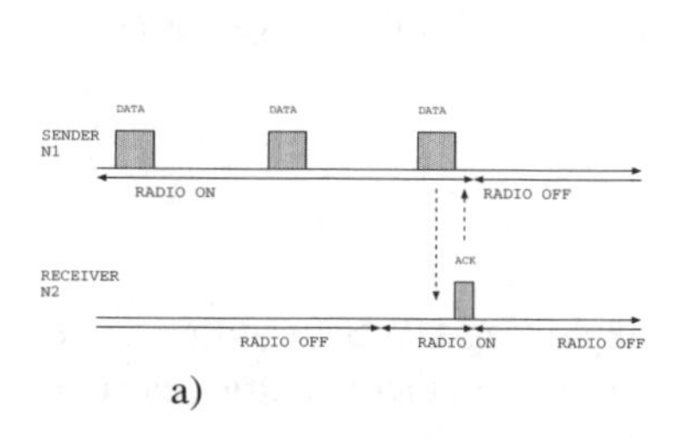

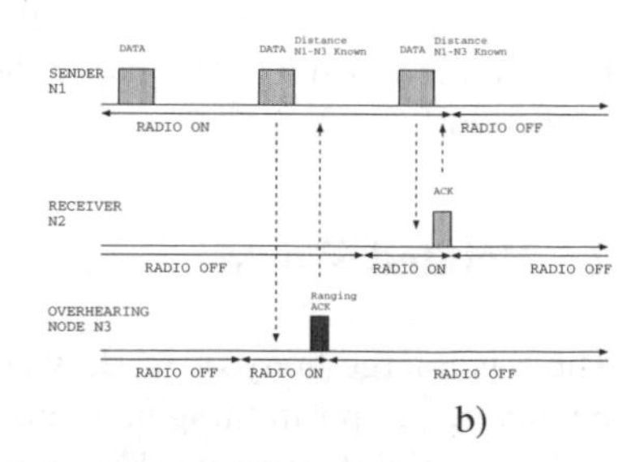

Fig. 1. FrameComm and FrameCommDM comunication mechanism

3.1 FrameComm

FrameComm, like many wireless contention based MAC protocols, performs duty cycling of node transceivers. To ensure that rendezvous between transceivers occur, Frame-Comm deploys a method in which a trail of identical packets of data, called framelets, is transmitted by the sender with gaps between each. The receiver sends an acknowledgement to the source after successfully receiving a framelet. Upon the reception of this acknowledgement, the sender may then cease sending and yield control of the channel (See Figure 1).

Assumptions and Definitions : It is assumed that the clocks of the transmitter and receiver operate at approximately the same rate. Note that this does not imply time or sleep cycle synchronization; rather the clock drift between any two nodes is insignificant over a short period. It is also assumed that a fixed rate radio duty cycle is used, i.e., each node periodically activates its radio for a fixed time interval to monitor activity in the channel. The duty cycle period is represented as $P = \triangle + \triangle_0$, where $\triangle$ is the time the radio remains active and $\triangle_0$ is the time the radio is in sleep mode. The duty cycle ratio is defined as:

$$DutyCycle = \frac{\triangle}{P} = \frac{D}{\triangle + \triangle_0} \tag{1}$$

Rendezvous using Framelets : Framelets are small, fixed-sized frames that can be transmitted at relatively high speeds. Successful duty cycle rendezvous require a sequence of identical frames to be repeatedly transmitted from the source node; each frame contains the entire payload of the intended message as depicted in Fig. 1. If the receiver captures one of these, the payload is delivered. The trail of framelets is defined by three parameters: Number of transmissions: n ; time between framelets: δ_0 ; framelet transmission time: δ .

To achieve successful rendezvous a relationship must be established between the parameters $\triangle$, $\triangle_0$, n, δ, and δ_0. First, the listening phase of the duty cycle $\triangle$ must be such that: $\triangle \geq 2 \cdot \delta + \delta_0$. This ensures that at least one full framelet will be intercepted during a listen phase. Furthermore, to ensure overlap between transmission and listening activities, the number of retransmissions n needs to comply with the following inequality when $\triangle_0 > 0 : n \geq [\triangle_0 + 2 \cdot \delta + \delta_0/(\delta + \delta_0)]$. This ensures that a framelet trail is sufficiently long enough to guarantee rendezvous with the listening phase of the receiver, and ensures that at least one framelet can be correctly received. The duration of $\triangle$ determines message delay, throughput and energy savings.

Message Acknowledgments : Between framelet transmissions, the source node switches its radio to a listening state. Upon successful reception of a frame at the destination node, this receiving node should respond with an acknowledgement transmitted during the framelet transmission gaps δ_0. After reception of this acknowledgment the sender should terminate transmission of its framelet trail as communication has been successful. The use of acknowledgments reduces the amount of framelets needed for each transmission, and as a result, transmissions will occupy the channel for a shorter period of time, reducing contention whilst increasing throughput and energy efficiency.

3.2 FrameCommDM

The basic principle of FrameComm is ideally suited for the integration of positioning functions. The method of exchanging packets and acknowledgements mirrors that of two-way-ranging methods used to determine the round-trip-time, and ultimately the TOF of signals. If the sender records the time of transmission of its last framelet, and the time upon receiving its acknowledgement, the distance between nodes can be determined. Furthermore, a sender may derive not only the distance to its intended recipient, but potentially the distance to any node within transmission range. During the exchange of framelets between the sender and receiver, a third node may enter its listening period, overhear a framelet and respond with a so called ranging acknowledgement (See Figure 1.b.).

Basic Ranging : To determine the distance between two communicating nodes the time-of-flight (TOF) of exchanged signals needs to be measured. To avoid the need of tight clock synchronization between both nodes two-way-ranging can be performed using the existing FrameComm data exchange. The sender of a message keeps track of the time t_t when a framelet is transmitted. If an acknowledgement is received, its arrival time t_a is recorded. The TOF can be determined using t_t and t_a if the processing time t_p at the message receiver is known. The processing time t_p is the time required by the message receiver to respond with an acknowledgement to the received framelet. It is assumed that the processing time t_p is constant and thus known by the message transmitter. The TOF can be calculated as: $TOF = (t_t - t_a - t_p)/2$. The distance between the two nodes is proportional to the measured TOF.

It has to be noted that a transmitter of a message can determine the distance to the message receiver without consuming additional energy for ranging as existing messages are used. Likewise, network performance in terms of achievable throughput and message transfer delay is not degraded by introducing ranging.

Ranging Acknowledgements : The previously outlined basic ranging mechanism can be improved by introducing ranging acknowledgements. The improvement exploits the fact that nodes not directly involved in the message transport might overhear framelets.

During regular communication a source node will generate data and begin transmitting its framelet trail, and await an acknowledgement. It is possible for nodes whom the packet is not the intended recipient to overhear framelets of the transmission. Normally, a node overhearing a packet not addressed to it would simply ignore the received packet and enter its sleep cycle. However, to improve ranging we propose that a node sends a ranging acknowledgement packet before entering the sleep state. Thus, a sender of a message does not only obtain the distance to the communication partner, but will potentially also collect distance information to nodes overhearing the communication (See Figure 1.b.).

This ranging acknowledgement is not sent immediately after the framelet is received. The transmission of the ranging acknowledgement is delayed by a time δ_R which is greater than the time needed to transmit a message acknowledgement. Thus, collisions between ranging acknowledgements with the message acknowledgement are avoided (See Figure 1.b.). In some cases, ranging acknowledgements transmitted by several overhearing nodes in response to the same framelet might collide. However, this will only reduce the effectiveness of the positioning function of FrameComm but will not have an impact on message transmission or network performance.

Ranging acknowledgements are transmitted within the gaps of an existing framelet trail. Thus, the introduction of ranging acknowledgements has no immediate impact on the network performance in terms of message transfer delay or network throughput (See experimental evaluation in Section 5). Energy consumption of nodes may vary by the introduction of ranging acknowledgements as additional messages need to be transmitted. However, our experiments show that this variance is acceptably small.

4 Prototype Implementation

Our choice of an UWB transceiver type was determined by our needs for implementing FrameCommDM. More specific, the transceiver hardware must provide a programming interface which allows us to implement FrameCommDM. Furthermore, we took power consumption of available transceivers into account. A low power consumption is important to make the system viable for most WSN deployment scenarios. Of the possible candidate systems the Nanotron NA5TR1 [3] best fitted these requirements.

4.1 Prototype Platform

For the FrameCommDM implementation we used the nanoLOC development board which comprises an of ATMega 128L microcontroller with 128*kb* flash memory and 4*kb* of SRAM, driving a nanoLOC EVR module; the main components of which being the NA5TR1 transceiver. Although it would be possible to port a common sensor network operating systems such as TinyOS or Contiki to the nanoLOC platform, we decided to implemented our own simple OS solely for testing MAC protocol performance.

The Nanotron NA5TR1 transceiver uses chirp spread spectrum (CSS), which is included as an alternate physical layer specification in the IEEE 802.15.4a. CSS is similar to other spread spectrum techniques in that it uses the entire allocated bandwidth to transmit a signal, however, CSS uses Linear Frequency Modulation (LFM), called chirp pulses, which fill the allocated bandwidth over a predefined duration. This makes CSS modulation resilient against channel noise, and also robust against multipath signals, allowing for increased accurate estimation of the Time-Of-Arrival (TOA) of the Line-Of-Sight (LOS) signal, which therefore results in greater accuracy of range estimations. The transceiver operates in the $2.45GHz$ ISM band at programmable data rates of between $125Kbps$ and $2Mbps$, with typical current consumptions of $35mA$ while transmitting, $33mA$ in a receiving state, and $2\mu A$ in a shutdown state. These figures compare to those of the Texas Instruments CC2420 transceiver [9] commonly used in WSN platforms as $17.4mA$ transmitting, $19.7mA$ receiving, and $20\mu A$ when shutdown.

4.2 Nanotron Ranging API

To facilitate distance estimations the transceiver hardware provides two types of ranging, defined as *normal* and as *fast* ranging. Nanotron's normal ranging technique allows both participating nodes to perform range estimations and the initiating node to average the estimations for increased accuracy (see [3]). Nanotron's fast ranging mode is similar to that of the two-way ranging method of range estimation, only one round trip measurement is used for range estimation. It has to be noted however that 2 acknowledgments are used. The first is a hardware generated acknowledgement (also called ranging pulse), having a fixed payload which cannot carry user data. If the receiver intends to report data in the acknowledgement a second data transmission must follow the ranging pulse (see Figure 2). Essentially the acknowledgement is split into two packet transmissions. Despite the additional ranging pulse, the fast ranging implementation closely resembles that of the packet exchange of the FrameCommDM protocol. Therefore, the fast ranging capability of the Nanotron hardware was used to implement FrameCommDM.

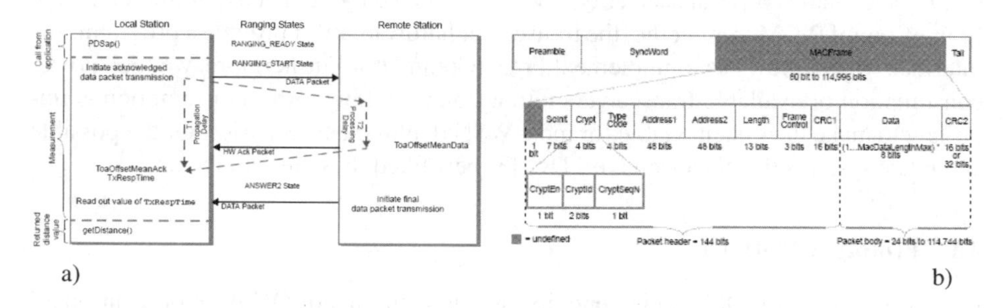

Fig. 2. *a)* Fast Ranging. Node 1 Generates a packet, sends to Node 2. Node 2 returns a hardware generated ack, followed by a data packet to transmit user data. Node 1 estimates distance to Node 2; b) Packet format the Nanotron NA5TR1.

The data packet format of the transceiver consists of a 30*bit* preamble, a 64*bit* sync word, a 4*bit* tail, and a MAC frame with a header size of 176*bits* and a maximum data payload field size of 8192*bytes* (see Figure 2). This is an increase of 50% in header size when compared to the frame format of the CC2440; assuming the same 48*bit* addressing scheme.

Obviously the efficiency of our FrameCommDM implementation is limited due to hardware constraints. Packets are relatively large and an additional short ranging pulse has to be transmitted for each message exchange. However, these issues could be addressed with hardware re-design. The header size could be reduced and the ranging pulse could carry user data.

4.3 FrameCommDM Implementation

Parameter Settings: The implementation values for various FrameComm parameters, as outlined in Section 3, have to be selected. The maximum framelet size is 578*bit* in length, consisting of a 274*bit* of frame header, 136*bit* of FrameComm header, and a maximum payload size of a simple 8*bit* sensor reading, and four times a structure of 8*bit* node ID and 32*bit* range estimation. This allows each node to forward a sensor reading with attached range estimation of a maximum of four neighbours to the sink for each message. We assume the sink is collecting ranging information and executes a positioning algorithm. This equates to a Framelet transmission time $\delta = 2.3ms$. We implement a 2% duty cycle using a framelet transmission period of $P = 600ms$ with a listen period $\triangle = 12ms$ and a sleep period $\triangle_0 = 588ms$. To satisfy the FrameComm specifications that $\triangle \geq 2 \cdot \delta + \delta_0$, where δ_0 is the time between framelets, and δ the framelet transmission time, we determine the Framlett interval to be $\delta_0 \leq 7.37ms$. Given that it takes under $2ms$ to generate a data ack of the required size of $410bit$, we can safely assume that after $\delta_R = 4ms$, overhearing nodes may choose to send ranging acknowledgements which will not collide with the data acknowledgements, and still arrive in adequate time to be received and processed before the next framelet transmission.

Fast Ranging: The implementation uses nanotron's fast ranging mechanism. Thus, an acknowledgement is split in two parts: ranging pulse and acknowledgement (see Figure 3). As previously explained, δ_R prevents collision of ranging acknowledgement and data acknowledgement, however, the collision of ranging pulses cannot be avoided with the existing hardware. Therefore a ranging measurement must be discarded if ranging acknowledgement and data acknowledgement are received after framelet transmission as it indicates a ranging pulse collision.

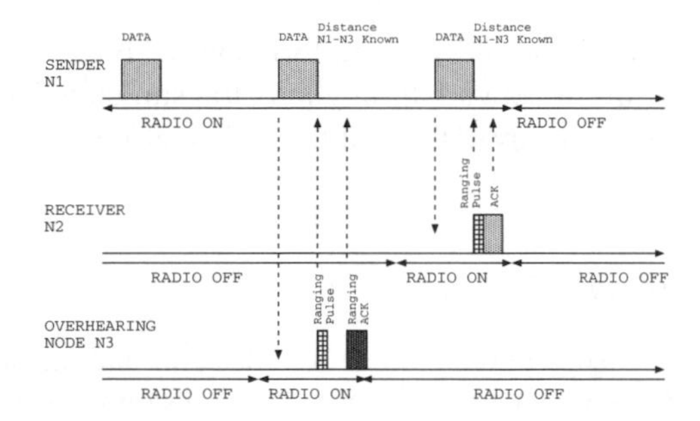

Fig. 3. FrameCommDM implementation on the Nanotron NA5TR1 using fast ranging

MAC Address Handling: For the NA5TR1 transceivers to facilitate the fast ranging mode, a sender must initiate a transmission with the $3bit$ Frame Control Field of the MAC frame set to $0x04$ and the frame must be addressed to a unicast address. A transceiver which is listening within communication range will check the frame control flag and reply with a ranging pulse to complete the TOF measurement, should this node have the destination address. The MAC layer may then retrieve the received data from the transceiver and return a ranging acknowledgement following the ranging pulse.

To implement our concept of ranging acknowledgements on the Nanotron hardware, all receiving nodes must have the same physical layer address, otherwise the transceiver would determine a packet not to be destined for this node, and therefore not send a ranging pulse. Ranging cannot be successful on this hardware if a node sends to its own physical address. Therefore, after a node has sampled the channel, before sending its framelet trail, it first sets its physical layer address to a different address to that of the global receive address. Once communication is successful, the node switches its physical layer address back to the predetermined global receive address. To uniquely identify a node sending an acknowledgement the payload uses the previously mentioned FrameComm header. Overhearing nodes can respond with a hardware pulse, followed by a ranging acknowledgement containing the node's FrameComm address.

4.4 Findings

The given hardware features of the NA5TR1 limit efficient implementation of FrameCommDM. However, these inefficiencies can be addressed by redesigning the NA5TR1 hardware. In particular, we recommend the following changes to allow efficient combined communication and positioning: (i) Ranging pulses should be able to carry user data. This would remove the necessity to send ranging pulses followed by a separate acknowledgement. (ii) It should be possible to define a ranging pulse transmission delay in order to avoid acknowledgement collisions. (iii) The transceiver should be able to respond to ranging requests not addressed to the transceiver. This would allow overhearing nodes to respond to ranging requests without MAC address modification.

5 Evaluation

The evaluation of this work considers the performance of FrameComm against that of FrameCommDM, in terms of network throughput, transmission delay and energy consumption. Furthermore, we analyze the ability of FrameCommDM to collect ranging measurements.

5.1 Experimental Setup

The system is evaluated using 5 nodes where 3 nodes are sending data via a forwarding node to a sink node. The sink nodes transceiver is always on while all other nodes use a 2% duty-cycle. The three leaf nodes and the forwarding node generate traffic destined for the sink. The setup is tested using different message generation frequencies λ as a parameter to vary traffic load. Each node generates messages every $1/\lambda$ ($1s \leq \lambda \leq 20s$) with an induced random jitter of $\pm 100ms$; thus nodes do not generate messages synchronously. Each experiment run is 5 minutes in length, and is repeated three times.

For each experiment nodes are configured with a buffer size of $b = 15$. A node can hold 15 messages in its forwarding buffer in addition to one that might currently be in the sending buffer. Messages are placed in this buffer when generated or received for forwarding. This value has been used to ensure messages are not dropped due to lack of buffer space in our experimental setup. Messages remain in the send buffer and will retransmit indefinitely until successfully acknowledged. If a node has multiple messages in its buffer, it sets a flag in packets of a framelet trail to indicate that it has multiple messages to send to the same receiver. When the destination node acknowledges the reception of the first packet, it examines this flag and stays awake to receive the remaining packets.

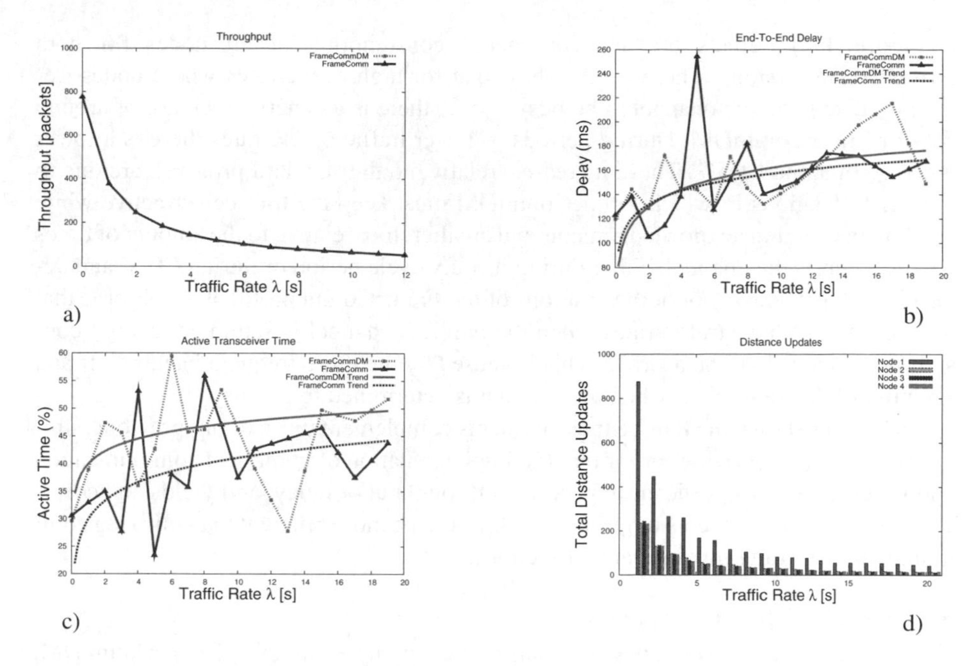

Fig. 4. a) Average throughput; b) Average message end-to-end delay; c) Average transceiver duty cycle of all nodes; d) Average number of distance measurements obtained by forwarding node and leaf nodes during experiment.

5.2 Communication Performance

To quantify the cost of implementing ranging into the FrameComm protocol, we use the experiment setup to determine throughput, delay and energy costs for FrameComm and FrameCommDM. Throughput is measured by determining the number of messages received at the sink over the duration of an experiment run. Delay is the end-to-end message delay for each message successfully delivered to the sink. Energy cost is determined by recording the time the transceiver of a node is active (listening, receiving, sending) during the experiment run.

Figure 4 a) shows the achieved throughput for FrameComm and FrameCommDM. From this figure it can be determined that the ranging enhancements introduced in FrameCommDM have no significant impact upon throughput. From the data collected it is found that throughput of FrameCommDM is reduced by a maximum of 0.9% and an average of 0.2%.

The results for the average delay for all nodes are shown in Figure 4 b). These show that, despite anomalous fluctuations in the real world results, the average increase in delay of adding ranging to the standard FrameComm protocol is a maximum of 4*ms*, however FrameCommDM can frequently be seen to yield lower delays than that of the standard protocol. This trend shows that the implementation of FrameCommDM has little impact upon message delay. We believe that prolonged testing in controlled environments will reduce the impact anomalies seen here to produce results closely resembling those of the trend lines shown.

Figure 4 c) depicts the average energy consumption of all nodes for both FrameComm versions. These results show that for high traffic rates where nodes frequently determine the channel to be busy $5 \geq \lambda$, there is an energy increase of around 5% for FrameCommDM. During periods of lower traffic $6 \leq \lambda$ rates there is a lower increase of around 1.5%. These increases, relative to the standard protocol, are due to the small durations in which FrameCommDM must keep the transceiver active while sending the ranging acknowledgements and are therefore relative to the amount of times a node samples the channel, either during its duty cycle period or during CCA, and determines it to be busy. For both variations of the FrameComm protocol we observe that during times of high traffic rates when the wireless channel is saturated, energy consumption is less than the average. This is caused by the more frequent backing off and sleeping of the transceiver when the channel is determined to be busy.

The results shown in all three measurements complement those of our previous, simulated work. This provides proof that the implementation of combined communication and ranging has no significant impact upon throughput or delay, and yields an acceptable increase in energy consumption, which would be more efficient than utilizing communication and ranging as separate functions.

5.3 Ranging Update Frequency

The evaluation of this work does not examine the ranging accuracy of FrameCommDM, as this is determined by the accuracy of range estimations of the NA5TR1 hardware. In accordance with the NA5TR1 data sheet [3], the accuracy of this particular transceiver is claimed to be within $\pm 2m$ indoor, and $\pm 1m$ in open space.

The evaluation of ranging of this work is to determine how frequently nodes estimate the distance to neighbouring nodes. For each new message generated, the node appends the data of its ranging table to the message payload. This data can then be examined by each intermediate node to update its local ranging table, and examined by the sink when the message reaches this destination. It has to be noted that during the experiments the sink does not respond with ranging acknowledgements. As the sink does not duty cycle the transceiver it would respond to every single message in the network with a ranging ack which is unnecessary.

Figure 4 d) portrays the total number of range measurements determined by each node. As expected, when traffic rates are high, nodes receive more frequent range estimations in the form of acknowledgements from their destination node for each message. Nodes also receive more overhearing acknowledgements during times of high traffic, as neighboring nodes more frequently sample the channel, and upon determining it to be busy, send ranging acknowledgements. Intermediary nodes, such as the forwarding node for this topology, receive more ranging estimations. These nodes not only gather range estimations from their generated messages and subsequent ranging acknowledgements, but also from the acknowledgements of the packets which they forward and from the ranging acknowledgments for these.

6 Conclusion

We have demonstrated that combined positioning and communication is implementable within low power MAC protocols, if the sensor platform provides a state of the art communication transceiver. We have found that implementing ranging into the FrameComm protocol has insignificant impact upon network performance, and that we can transport ranging data within the network at no extra cost. We have also shown how currently provided hardware interfaces of transceivers such as the NA5TR1 should be altered to support combined ranging and positioning more efficiently.

We believe that MAC protocol modifications as described in the paper can be applied to most power efficient sensor network MAC protocols. Thus, most existing MAC protocols can be augmented to include efficient positioning services if adapted to new transceiver hardware.

References

1. Patwari, N., Ash, J.N., Kyperountas, S., Hero, A.O., Moses, R.L., Correal, N.S.: Locating the nodes. IEEE Signal Processing Magazine 22(4), 54–69 (2005)
2. IEEE 802.15 WPAN Low Rate Alternative PHY Task Group 4a (TG4a), http://ieee802.org/15/pub/TG4a.html.
3. Nanotron nanoLOC TRX Data Sheet (2007), http://www.nanotron.com
4. Benson, J., O'Donovan, T., Roedig, U., Sreenan, C.: Opportunistic Aggregation over Duty Cycled Communications in Wireless Sensor Networks. In: Proceedings of the IPSN Track on Sensor Platform, Tools and Design Methods for Networked Embedded Systems (IPSN2008/SPOTS2008), April 2008, IEEE Computer Society Press, St. Louis (2008)
5. Correal, N.S., Kyperountas, S., Shi, Q., Welborn, M.: An uwb relative location system. In: IEEE Conference on Ultra Wideband Systems and Technologies, pp. 394–397 (2003)
6. Cheong, P., Oppermann, I.: An Energy-Efficient Positioning-Enabled MAC Protocol (PMAC) for UWB Sensor Networks. In: Proceedings of IST Mobile and Wireless Communications Summit, Dresden, Germany, pp. 95–107 (June 2005)
7. Polastre, J., Hill, J., Culler, D.: Versatile low power media access for wireless sensor networks. In: Proceedings of the 2nd International Conference on Embedded Networked Sensor Systems, SenSys 2004, pp. 95–107. ACM Press, New York (2004)
8. Buettner, M., Yee, G.V., Anderson, E., Han, R.: X-mac: a short preamble mac protocol for duty-cycled wireless sensor networks. In: Proceedings of the 4th International Conference on Embedded Networked Sensor Systems, SenSys 2006, pp. 307–320. ACM Press, New York (2006)
9. Chipcon Products From Texas Instruments, CC2420 - Data sheet, http://focus.ti.com/lit/ds/symlink/cc2420.pdf
10. Alcock, P., Roedig, U., Hazas, M.: Combining Positioning and Communication Using UWB Transceivers. In: Krishnamachari, B., Suri, S., Heinzelman, W., Mitra, U. (eds.) DCOSS 2009. LNCS, vol. 5516, pp. 329–342. Springer, Heidelberg (2009)
11. PulsOn 200 Evaluation Kit from Time Domain, PulsOn P220 - Evaluation Kit, http://www.timedomain.com/products/P220aEVK.pdf

SPIDA: A Direction-Finding Antenna for Wireless Sensor Networks

Martin Nilsson

Swedish Institute of Computer Science (SICS),
Box 1263, SE-164 29 Kista, Sweden

Abstract. This paper presents the design, signal processing, and field measurements of SPIDA, a direction-finding antenna for the 2.4 GHz ISM band, intended for both communication and localization in wireless sensor networks. The main design goals for the antenna were small size, low production cost, low power consumption, low signal processing requirements, and low interfacing complexity. The most expensive part of SPIDA is its SMA connector. The RF-stage power consumption is the same as for a whip antenna. The angle-of-arrival can be computed from received-power measurements through a simple formula using on the order of ten multiplications. Controlling the direction of the antenna by a microprocessor requires only a pair of digital output pins. When field tested with the TI CC2500 radio chip, the RMS error for the uncalibrated antenna was less than 12° up to 100 m distance, covering nearly the full receiving range of the antenna at 1 mW transmitter output power. A distinguishing feature of the SPIDA antenna is the absence of side lobes, despite using a manufacturing-friendly and cost-conscious sparse ground plane.

Keywords: direction finding, antenna, parasitic elements, angle of arrival, localization, wireless sensor network.

1 Introduction

There are many potential uses for directional antennas in wireless sensor networks (WSN). For instance, the directivity can be used for economizing transmitted power, and it can also be used for localization, where a useful piece of information is the angle of arrival (AOA, a.k.a. direction-of-arrival, DOA) of a transmission. Many direction finding (DF) antennas have been proposed, including loop, Adcock-pair, pseudo-Doppler, and phased-array antennas [1, 2, 3], but they often have properties unsuitable for WSN applications, which demand small and inexpensive antennas, but are not allowed any complex circuitry nor computationally heavy signal processing.

In this paper, we describe SPIDA, *Sics Parasitic Interference Directional Antenna*, which satisfies the basic requirements of electronically steerable directional antennas for WSN (fig. 1). SPIDA was primarily developed for localization applications (i.e. concurrent communication and AOA measurements), but it can also be used for

P.J. Marron et al. (Eds.): REALWSN 2010, LNCS 6511, pp. 138–145, 2010.

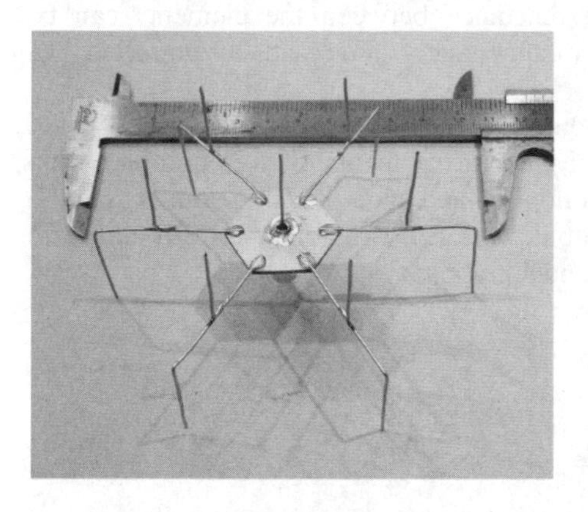

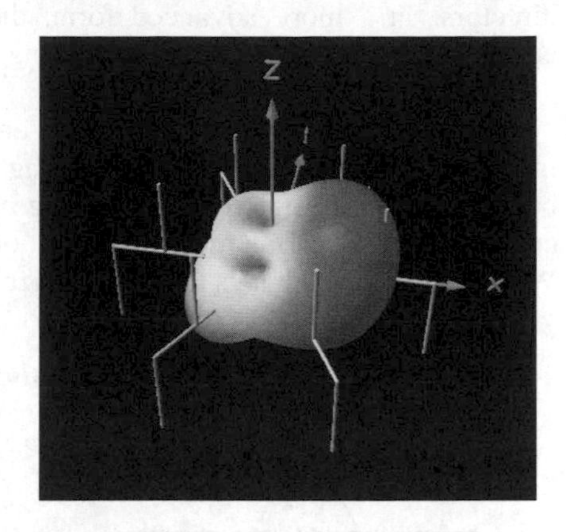

Fig. 1. (Left). The SPIDA antenna. In this setup, the rightmost parasitic element is isolated, while the others are grounded. The central element is a conventional monopole. **(Right).** The main lobe of SPIDA is quite smooth, as computed by 4Nec2 simulation.

enhancing transmission and reception in software-controllable directions. This paper focuses on the direction finding aspects of the antenna.

Radio direction finding antennas can compute AOA by measuring either the amplitude (power) of the incoming signal, or the phase (or both). An advantage of measuring the phase is that it is less affected by noise, compared to amplitude. However, RF phase measurement needs more advanced electronics with a highly stable timebase. Since this is presently difficult to achieve for an inexpensive WSN node, we have restricted the scope to amplitude measurements.

This paper is organized as follows: In the next section, we describe the design of SPIDA. In the following section, we describe how AOA data can be extracted from RSSI readings. In section 4, we describe the experimental setup and the measurement results. Section 5 concludes the paper.

2 The SPIDA Design

One of the major requirements of an antenna for WSN is small size. A fundamental fact of antenna theory says that an antenna cannot be much smaller in diameter than the wavelength [4]. If smaller, the efficiency drops dramatically. Since the 2.4-GHz ISM band can probably be considered a representative standard frequency for WSN applications, corresponding to a wavelength λ of approximately 120 mm, we can expect the antenna to be of roughly this size as well.

SPIDA, first conceived in [3], is a kind of *Electronically-Switched Parasitic-Element* (ESPE) antenna.. The ESPE principle was first published in 1979 [5], and has subsequently been developed further [6, 7]. An ESPE antenna consists of a central monopole, surrounded by a number of monopole-like parasitic elements spaced approximately $\lambda/4$ apart. In its simplest form, the parasitic elements are switched between ground, when they operate as reflectors, and isolation, when they operate as directors. In a more advanced form, the reactance between the elements can be controlled. This can be done by biasing a capacitance diode with a controlled DC voltage.

An attractive property of the ESPE antenna is that the parasitic elements are not involved in the active RF chain. The ring of parasitic elements on a ground plane can be added as a sleeve around an existing monopole, and each parasitic element can be controlled by a digital microprocessor output. The ESPE antenna is well suited to WSN applications, thanks to its small size and simplicity.

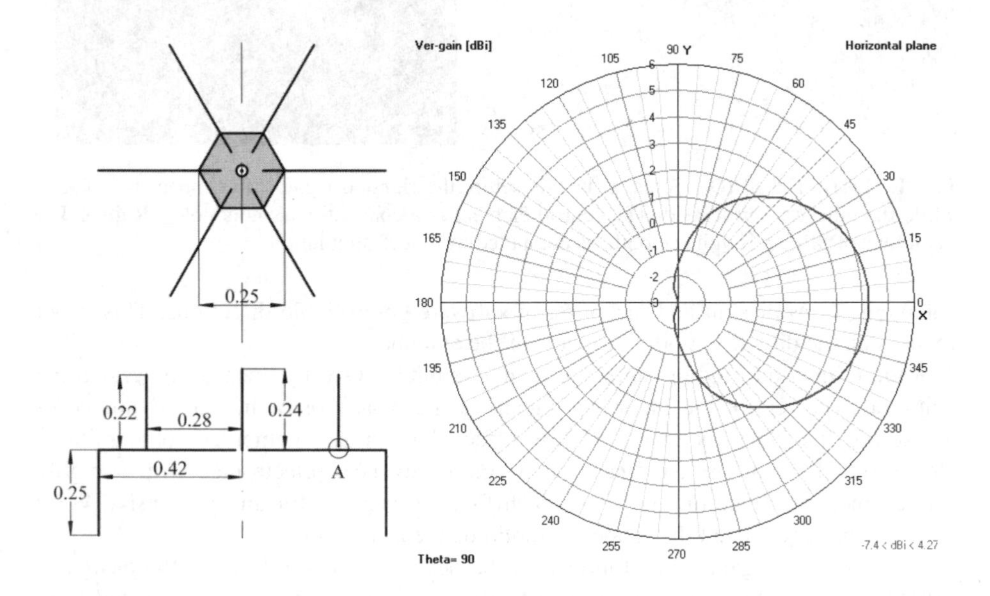

Fig. 2. (Left). Dimensions of SPIDA. Measures are in wavelengths. The directivity is controlled by isolating one element (A) in the preferred direction. **(Right).** Horizontal plane gain pattern computed by 4Nec2. The pattern has the desirable rounded shape.

Physical construction. The main distinguishing feature of SPIDA is its *sparse ground plane*. After extensive simulations it was found that this sparse ground plane gives the antenna a smooth main lobe without significant side lobes, which seems counter-intuitive from the viewpoint of classical antenna design. The absence of side lobes is highly desirable for the purpose of efficiently extracting AOA data from amplitude (RSSI) readings (cf. section 3 below).

SPIDA consists of a 3cm-diameter hexagonal disc made of ordinary copper-clad FR-4 circuit board, on which six legs made of standard 1-mm copper wire have been soldered as ground plane. No solid ground skirt is used. An SMA-connector is mounted centrally on the disc. The WSN node can be attached directly to this connector, eliminating the feed line.

The parts cost for SPIDA is dominated by the SMA connector (~10 USD in single quantities). The antenna can be built in a short time, and can be easily tuned, thanks to the active and passive elements being ordinary copper wires, which can be bent and cut with good precision after soldering. The antenna does not require any additional amplifier beyond that needed by the central monopole. The parasitic elements can be switched at low frequency, without consuming DC power.

Nominal characteristics. The antenna was initially developed and optimized using the 4Nec2 antenna simulation package [8, 9]. The nominal performance, as indicated by this program, is an antenna gain of approximately 4 dBi and a front-to-back ratio of 11 dB. The main lobe is smooth (fig. 1, right) and the gain (in dB) in the horizontal plane is nearly an offset circle (fig. 2, right). Since the antenna is supposed to be useful throughout the 2.4-GHz band (2400-2480 MHz), the antenna impedance varies, and it is not possible to design a fixed impedance matching circuit perfect throughout this frequency range. On the other hand, since the distance between the antenna and the RF output amplifier is normally much shorter than the wavelength in a WSN node, the need for precise impedance matching is relaxed.

Interfacing. The SPIDA antenna measured uses a fixed directing element and fixed reflecting elements, but normally, the elements are switched between ground and isolation under program control.

An advantage of ESPE antennas greatly simplifying implementation is that the control of this switch is not part of the RF path, and can be done arbitrarily slowly. There are three common methods for implementing the switch [10]. The traditional method uses PIN diodes, consuming a bias current, and a fair number of additional components. A more recent approach uses GaAs pHEMT transistors, requiring fewer discrete components, but such transistors are relatively expensive. Recently, however, inexpensive CMOS RF switches have become available, e.g. ADG902 from Analog Devices and BGS12A from Infineon, rendering the interfacing of SPIDA close to trivial. The CMOS switches consume practically no quiescent DC current and require no additional components. An interface for SPIDA using ADG902 was built and evaluated by Öström et al. [11].

In order to control the switches from a microprocessor, one digital output is required for each switch. If an external shift register is used, a pair of digital outputs suffices, one for data and one for clock output.

3 Signal Processing

A WSN node possesses only limited resources in terms of time and power for signal processing in order to extract AOA data from a transmission. For this reason, it is crucially important that the lobes can be formed so that efficient signal processing algorithms can be applied. The ideal pattern in this respect is an offset circle, since the received power then approximates a cosine as a function of the AOA, and the AOA can be estimated as the phase of that cosine. For this case, there is a highly efficient computational method, which extracts the phase of a sine wave with a known frequency [12]. Suppose that P_k, $0 \le k < N(=6)$ is the power measured when parasitic element k is isolated (i.e. becomes a director). Let

$$S = \sum_{k=0}^{N-1} \sin\left(\frac{2\pi k}{N}\right) P_k$$
$$C = \sum_{k=0}^{N-1} \cos\left(\frac{2\pi k}{N}\right) P_k \qquad (1)$$
$$A = \sqrt{S^2 + C^2}$$

Then, for the estimate α of AOA,

$$\sin\alpha = C / A$$
$$\cos\alpha = -S / A \qquad (2)$$

Fig. 3. Measurement setup. The receiver is in picture. The transmitter is located in the upper right corner, at the street sign at the end of the road.

4 Results

4.1 Test Setup

We measured SPIDA in outdoor and indoor experiments. For measurements, we used Texas Instruments' CC2500DK development kit [13]. This kit contains two microcontroller boards and two CC2500 radio daughter boards, together with software that can transmit test packets, and collect RSSI measurements from the CC2500 chips. SPIDA was attached to one radio board used as a receiver, while a conventional monopole was attached to the other board used as transmitter. The test software SmartRF Studio 7 was executed on two PC laptops, each connected to its own microcontroller board.

The transmitter was placed on a chair, giving the antenna a height approximately 0.5 m above ground. The transmitter was set up to continuously send packets of 24 bytes including preamble, at 2.4 kBaud on 2440 MHz. Other frequencies were also tested, but the choice of frequency appeared not to affect measurements. The transmitter output power was 1 mW. The receiver was placed on a rotating table (fig. 3), which was then placed on a chair at about the same height as the transmitter. The rotating table was rotated manually, and the average RSSI for 16 packets was recorded at 22 positions around the table. For each distance, three such series were measured and the median of the readings were taken as the final RSSI measurement. The reference direction was set by eyesight. The antenna was not calibrated or tuned in any way before measurements.

We performed a total of four experiments, measuring outdoors at distances of 20m, 50m, and 100m between transmitter and receiver, and indoors at a distance of 7m. At 100m distance, the received power was nearly down to the noise floor (~ -105 dBm) and there was considerable packet loss.

4.2 Measurement Results

The AOA estimates showed reasonably linear responses for all four experiments (fig. 4, top). The RMS error was less than 12° in all cases (table 1). The performance appears to deteriorate slowly with distance, but remains reasonable throughout the receiving range of around 100m (fig. 4, bottom). Part of the explanation for the offset error is probably the manual zero adjustment, which was not very precise.

Table 1. RMS error and received power for the test settings

Setting	AOA RMS error	Average received power
Outdoor, 20m	11.97°	-70 dBm
Outdoor, 50m	6.14°	-86 dBm
Outdoor, 100m	10.49°	-99 dBm
Indoor, 7m	11.67°	-63 dBm

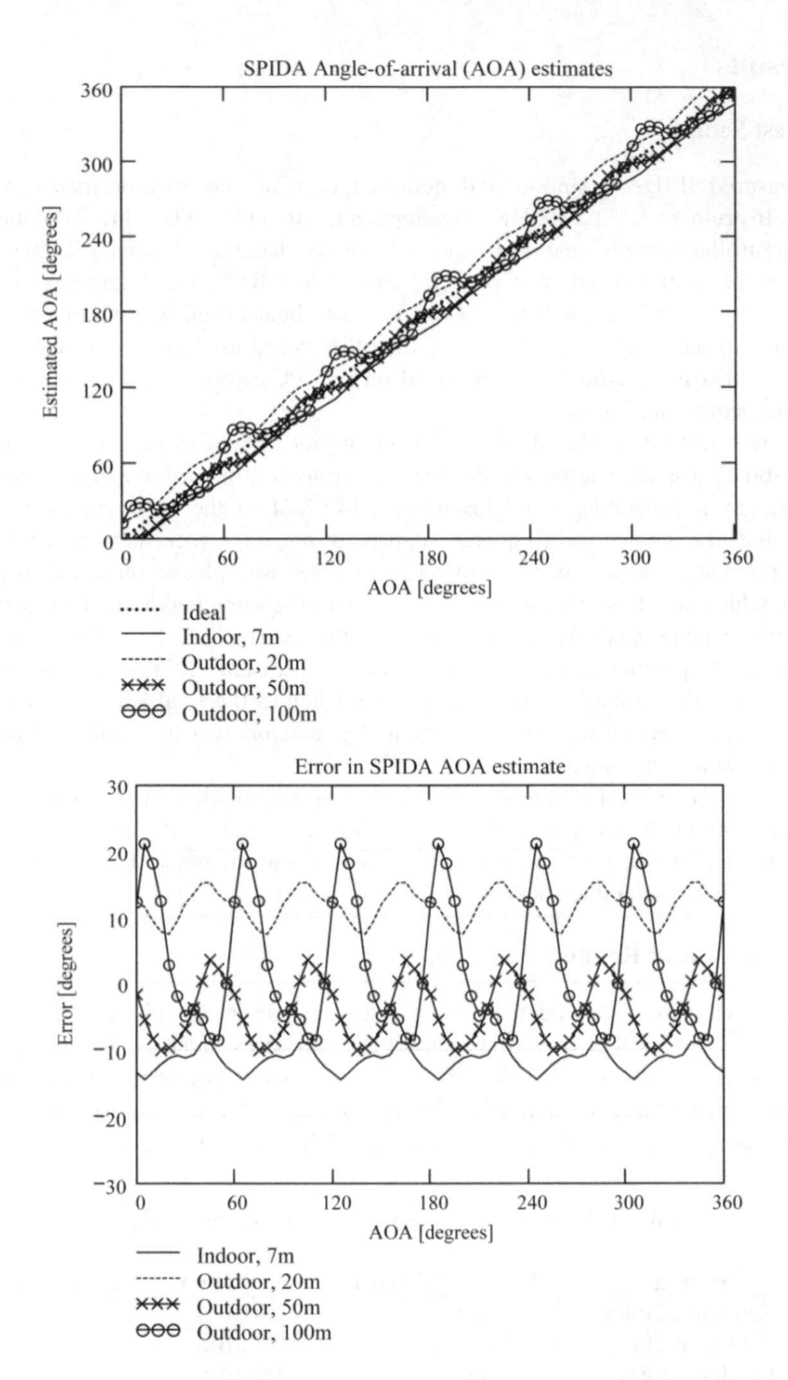

Fig. 4. (Top) AOA values show linear behavior for all distances. **(Bottom)** Error deteriorates as distance increases, but remains reasonable throughout receiving range.

5 Conclusions

We did not make great efforts to find noise-free locations for measurements, but the indoor measurements did require an adjustment in order to avoid a noisy location in the neighborhood of a WLAN access point. In general, one should expect indoor AOA measurements to be difficult and require redundancy. We first tried outdoor measurements in a soccer field, which failed, probably because a surrounding metal fence created a nearly homogeneous radiation pattern. Anyway, the RMS error of 12°, which must be considered low for the circumstances, indicates that the simple SPIDA approach may indeed be viable in many situations.

Acknowledgments. This work was carried out within the SICS Center for Networked Systems, funded by VINNOVA, SSF, KKS, ABB, Ericsson, Saab Systems, TeliaSonera and T2Data.

References

1. American Radio Relay League: The ARRL Handbook for Radio Communications (2009) ISBN 0-87259-146-8
2. Carr, J.J.: Practical Antenna Handbook, 4th edn. McGraw-Hill, New York (2001) ISBN 0-07-137453-3
3. Nilsson, M.: Directional antennas for wireless sensor networks. In: Proc. 9th Scandinavian Workshop on Wireless Adhoc Networks (Adhoc 2009), Uppsala, Sweden, May 4-5 (2009)
4. Hansen, R.C.: Electrically Small, Superdirective, and Superconducting Antennas. John Wiley, Chichester (2006) ISBN 978-0-471-78255-1
5. Harrington, R.F.: Switched Parasitic Antennas for Cellular Communications. IEEE Trans. Antennas and Propagation AP-26(3), 390–395 (1978)
6. Thiel, D.V., Smith, S.: Reactively controlled directive arrays. Artec House, Inc., Boston (2002) ISBN 1-58053-154-7
7. Schlub, R., Lu, J., Ohira, T.: Seven-Element Ground Skirt Monopole ESPAR Antenna Design From a Genetic Algorithm and the Finite Element Method. IEEE Trans. Antennas and Propagation 51(11) (2003)
8. Voors, A.: 4Nec2: Nec-based antenna modeler and simulator (2010), `http://home.ict.nl/~arivoors/`
9. Burke G.J., Poggio, A.J.: Numerical Electromagnetics Code (NEC): Method of Moments.Lawrence Livermore Laboratories, USA. Technical Report UCID-18834 (1981)
10. RF CMOS SPDT Switches. Infineon Application Note No. 175 (2009)
11. Öström, E., Mottola, L., Voigt, T.: Evaluating an Electronically Switched Directional Antenna for Real-world Low-power Wireless Networks. In: Proc. Int. workshop Real-world wireless Sensor Networks, Colombo, Sri Lanka, December 17-19 (2010)
12. Rauch, L.L.: On Estimating the Phase of a Periodic Waveform in Additive Gaussian Noise - Part I. NASA DSN Progress report 42-45 (1978)
13. CC2500DK Development Kit User Manual 1.4 (Rev. C). Texas Instruments (2007)

Testing Selective Transmission with Low Power Listening

Morten Tranberg Hansen[1], Rocío Arroyo-Valles[2], and Jesús Cid-Sueiro*

[1] Aarhus University,
Aabogade, 34, 8200 Aarhus, Denmark
mth@cs.au.dk

[2] Universidad Carlos III de Madrid,
Av. de la Universidad, 30, 28911 Leganés-Madrid, Spain
{marrval, jcid}@tsc.uc3m.es

Abstract. Selective transmission policies allow nodes in a sensor network to autonomously decide between transmitting or discarding packets depending on the importance of the information content and the energetic cost of communications. The potential benefits of these policies depend on the capability of nodes to estimate its current energy consumption patterns. As a case study, this paper tests the performance of a particular selective transmission algorithm over a simple network using a low power listening MAC protocol on real sensor node hardware.

Keywords: selective transmission, implementation, energy-aware.

1 Introduction

To overtake the energy limitations in sensor networks, selective transmission strategies allow nodes to discard messages if the energy cost of transmission is not compensated by the importance of message content. This is used as a basis for censoring sensors in decentralized detection [1] and other schemes [5,2]. A common assumption in these proposals is that the energy consumption of the sensors and their battery level are known. Thus, free parameters of the selective algorithms can be directly related to the energy consumption [2] or depend on them [6]. Moreover, to increase the benefits of discarding messages, selective transmission strategies assume that the cost of transmission is much higher than that of reception. These assumptions are not trivially satisfied. The average power consumption of transmission and reception depends not only on protocol design at all levels of the communication stack but also on the specific network deployment. As a consequence, the configuration of the free parameters in selective transmission algorithms cannot be done in advance and hence adaptive techniques must be applied.

This paper analyzes the problem of estimating the free parameters of a selective transmitter by correctly estimating the energy spent in each of its respective

* This work was partially funded by projects TEC2008-01348 from the Spanish MCI, and SensoByg (http://sensobyg.dk/english) at Aarhus University.

P.J. Marron et al. (Eds.): REALWSN 2010, LNCS 6511, pp. 146–153, 2010.

states. Our work can be taken as a case study: we have tested the implementation of the selective transmission strategy proposed in [2] using TinyOS with its Low Power Listening (LPL) MAC layer running on a number of Tmote Skys (motes).

2 Selective Transmission

The selective transmission model in [2] defines the state of any given node at time k by two variables: the energy reserve, e_k, and the importance of the message to be transmitted, x_k. This importance value may reflect, e.g., the priority, the relevance or the quality of the information conveyed, and it is assumed to be provided by the application layer to the node carrying the message. Furthermore, we assume that the network routing algorithm (whatever it is) has defined a set of neighbors for each node, in such a way that any sensor node holding a message at time k has to make a decision about sending or not the message to one of its neighbors. The decision rule is a function of the node state.

We consider four energy expenses that deplete batteries: data collection by a sensing device, e_S, reception of a message from other node, e_R, message transmission, e_T and idle state e_I. After receiving (or sensing) a message, the battery level decreases e_R (or e_S) if it is not transmitted, and $e_R + e_T$ (or $e_S + e_T$) if it is transmitted. If no messages are received, the node consumes e_I. In general, e_S, e_R, e_T and e_I may depend on k.

A selective transmitter is allowed to discard low graded messages with the expectation of transmitting more important upcoming messages later. Optimal policies to maximize the expected importance sum of all messages transmitted during the node lifetime have been derived [2]. If the importance sequence, x_k, is stationary, it turns out that, for large values of the available energy, e_k, the optimal decision rule is: to transmit the message at time k iff the importance value is higher than some threshold, μ, which is given by the solution of

$$\mu = \rho \mathbb{E}\{\max\{x_k - \mu, 0\} | x_k > 0\}. \tag{1}$$

Though (1) cannot be solved in closed form, we derive the following stochastic rule which provides a simple estimate that converges to μ for large k,

$$\mu_k = \left(1 - \frac{1}{k}\right) \mu_{k-1} + \frac{\rho}{k} \cdot \max\{x_k - \mu_{k-1}, 0\}. \tag{2}$$

Parameter ρ depends on the average node statistics as

$$\rho = \frac{(1 - P_I)E_T}{P_I E_I + P_R E_R + P_S E_S}, \tag{3}$$

where $E_S = \mathbb{E}\{e_S\}$, $E_R = \mathbb{E}\{e_R\}$, $E_T = \mathbb{E}\{e_T\}$, $E_I = \mathbb{E}\{e_I\}$, and P_I, P_R and P_S are the probability of a node being in idle, *receive* or sensing mode, respectively[1]. The estimation of all these energetic statistics is discussed in the following.

[1] Actually, the model in [2] does assume $E_S = E_R$. We have generalized the model to assume different costs for sensing and receiving, which is more realistic.

3 Implementation

We implemented the selective transmission algoritm in TinyOS 2.x using its default CC2420 radio stack with its Carrier Sense Multiple Access (CSMA) channel access mechanism below its X-MAC [3] variant of a Low Power Listening (LPL) MAC layer. We used the component-based structure of TinyOS to add local modifications to the radio stack in order to support energy estimation according to the radio states and mote statistic according to the LPL layer states. Fig. 1 gives an overview of our implementation.

3.1 Energy Consumption

To measure the energy consumption online at a sensor we use software-based on-line estimation [4] based on the radio and sensor state to deplete a virtual battery. We neglect the energy consumed by the micro-controller. The energy measurements are based on the reported current consumptions from the data-sheets (19.6 mA during reception, 17.04 mA during transmission, 1 mA by the SHT11 Temperature sensor) multiplied by the amount of time spent in each state, measured with a 32 kHz timer. Battery is depleted according to the last state at every state change. The physical layer (PHY) of the radio stack was modified to report the off, receive, or transmit radio states to the energy measurement component (Fig. 1).

3.2 Mote Statistics

Using an asynchronous LPL MAC layer, sensor node lifetime is divided into three states: *idle*, *receive*, and *transmit*, from which we estimate the energy costs E_I, E_R, and E_T, and the probabilities P_I and P_R. These estimations are done in the mote statistic components based on the *idle*, *receive*, and *transmit* states together with all packet reception events and timer wake-ups reported from the MAC layer (see Fig. 1, which also shows how the mote statistics component uses the changing battery level from the energy measurement component).

Fig. 2 illustrates the definition of *idle*, *receive*, and *transmit* states reported from the MAC layer to the mote statistic component. Fig. 2(a) shows how the MAC layer handles receptions. After sleeping for a fixed period of time, it wakes up the radio, samples the channel for energy, and only stays awake if energy is detected. Otherwise, it is an *idle* state. Energy might be detected due to noise, which we also refer to as an *idle* state, or due to an actual packet reception, which

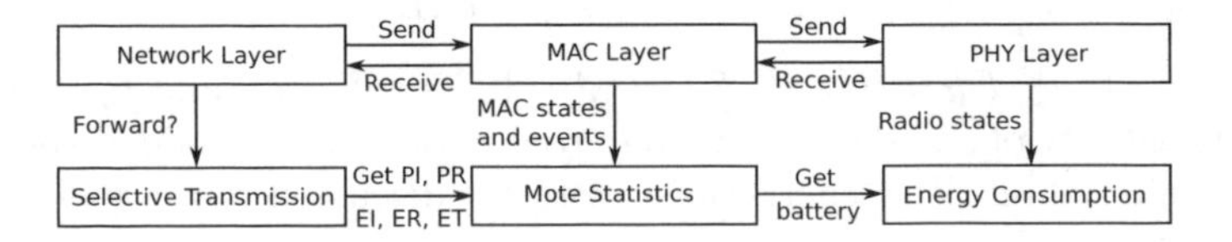

Fig. 1. Embedded Software Architecture

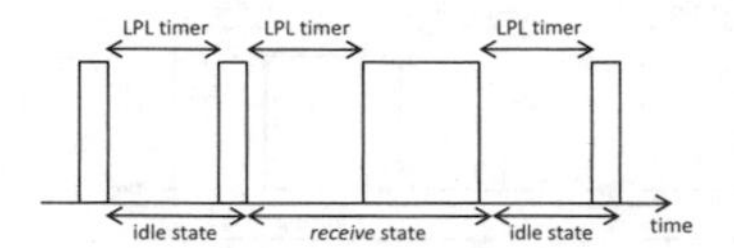

(a) *Idle* and *receive* states without the influence of overlapping *transmit* states.

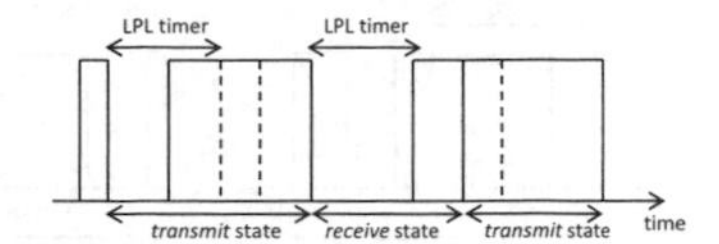

(b) Overlapping *transmit* states which takes over the *idle* or *receive* states.

Fig. 2. States of the LPL MAC layer. The dashed lines indicate what would have happened in case of no overlapping transmissions.

we refer to as a *receive* state. The length of a *receive* state varies with the number of received packets. To guarantee that a receiver is awake during transmission and acknowledges a successful reception, a transmitter repeatedly sends the same packet for a duration longer than the sleep interval. The transmission of a packet (referred to as *transmit* state) can interfere with the *idle* and *receive* states (see Fig. 2(b)). First, the transmission may happen during a sleep period, where the radio is then turned on and the next wake-up is ignored (an *overlap*). Second, the transmission may happen when the radio is already on so that the current *receive* (or *idle*) state is cutoff once the transmission starts (a *take-over*).

On top of normal packet receptions during a *receive* state and wake-ups, the MAC layer also reports free receptions. As packet transmission includes a waiting period for acknowledgments (ACKs) a sensor might receive a packet from another neighbor during transmission. We refer to these receptions as free in the sense that they do not imply any cost (the radio is in *receive* mode already).

Parameters E_I, E_R and E_T can be adaptively estimated as the sample averages of e_I, e_R and e_T, respectively, during the node lifetime. The idle state costs is the simplest case. If e_I is the energy consumption of the k-th idle state, the average idle energy is computed iteratively as $E_I(k) = \left(1 - \frac{1}{k}\right) E_I(k-1) + \frac{1}{k} e_I$. If m packets are received during a single *receive* state, then $E_R(k+m-1) = \left(1 - \frac{m}{k+m-1}\right) E_R(k-1) + \frac{1}{k+m-1} e_R$. The computation of the transmit state cost E_T is analogous to E_I, $E_T(k) = \left(1 - \frac{1}{k}\right) E_T(k-1) + \frac{1}{k} e_T$. However, the computation of e_T after each *transmit* state is quite involved. To understand it, note that e_T represents the cost overhead of deciding to transmit. If *transmit*, *receive* and *idle* states did not overlap, this would be equivalent to the energy of the *transmit* state. In general, however, this is not the case, and e_T must be computed as the difference between the cost of the *transmit* state (denoted as e_{T_r}) and the energy the node would have expended if the packet had been discarded. Therefore, a *transmit* state may take place in a non-interfering way (Fig. 3(a)), interfere with the *idle* and *receive* states by overlapping (Fig. 3(b)) or taking them over (3(c)). We analyze these three cases to compute e_T.

Case 1: Non-interfering transmissions (Fig. 3(a)). If no free receptions happen during the transmission, e_T is simply the cost of the *transmit* state e_{T_r}. However, if n free receptions happen, the overhead depends on the next state. If it is *idle*, it would have been a *receive* state in case we had not transmitted, so that

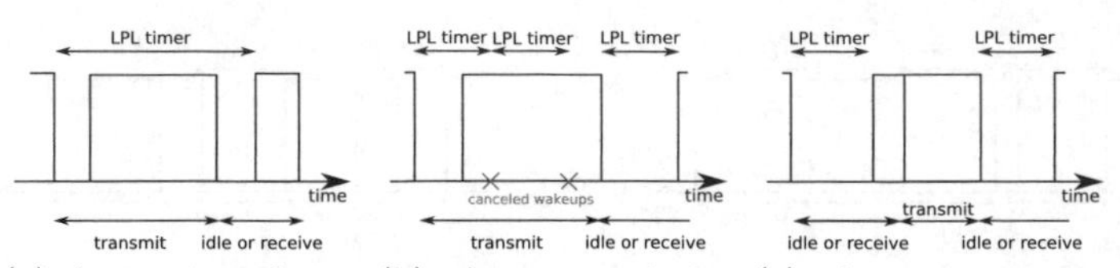

(a) A non-interfering transmission. (b) A transmission overlapping an *idle* or *receive* state. (c) A transmission taking over an *idle* or *receive* state.

Fig. 3. The three cases of a transmission

$e_T = e_{Tr} + E_I - nE_R$. But if it is a *receive* state receiving m packets, it would have been a *receive* state receiving $m+n$ packets in case we had not transmitted, and $e_T = e_{Tr} + mE_R - (m+n)E_R = e_{Tr} - nE_R$. We will later refer to the stored e_{Tr} and n as the pending transmission cost and pending free receptions.

Case 2: Overlapping transmissions (Fig. 3(b)). An overlapping transmission can be detected by the number of canceled wake-ups, w, according to the LPL MAC layer. If no free receptions happens, the cost needs to be compensated with w *idle* states. However, if n free receptions uniformly distributed during transmission happen, some of the *idle* states would have been turned into *receive* states. Then, the overhead is $e_T = e_{Tr} - nE_R - \max\{w - n, 0\} + E_I$.

Case 3: Taking-over transmissions (Fig. 3(c)). The transmission will shorten the *idle* or *receive* and then, it will be handled as a non-interfering transmission. Defining the measured energy cost of the taken over *idle* or *receive* state as d_I and d_R, respectively, the expected remaining energy cost of the taken over *idle* or *receive* state, if the transmission did not take place, is subtracted from the measured transmission cost e_{Tr}. Thus, the overhead of transmission is $e_T = e_{Tr} - (E_I - d_I)$ if the previous state was *idle* and $e_T = e_{Tr} - (E_R - d_R)$ if it was a *receive* state. In this case, we do not update E_I and E_R.

Furthermore, several successive (but still independent) transmissions may happen after each other. This has a consequence for the non-interfering transmission with free receptions as the overhead cannot be estimated until the next *idle* or *receive* state. In Fig. 4(a), the second transmission does not overlap with

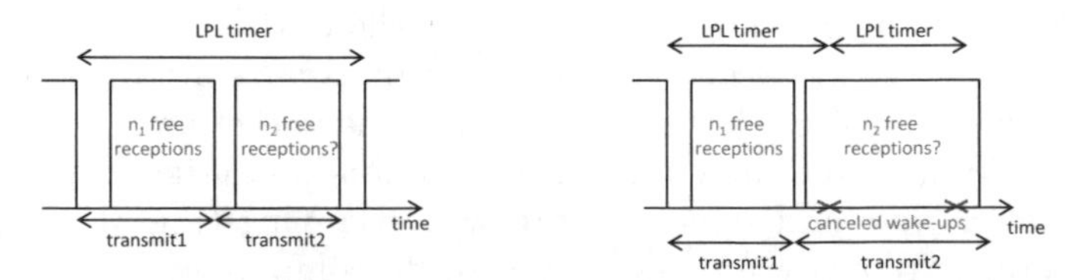

(a) Two non-interfering transmissions. (b) A transmission followed by another that overlaps with a number of wake-ups.

Fig. 4. Two different scenarios of a successive transmission following a non-interfering transmission with free receptions

any wake-ups. If it does not contain any free receptions, it can be handled as a non-interfering transmission. Instead, if the successive transmission contains free receptions, we have two pending transmissions whose overhead can not be estimated until the next *idle* or *receive* state. As E_T measures the average transmission overhead, we let the pending transmission costs and free receptions be an average of the two. In Fig. 4(b), the successive transmission does overlap with some wake-ups. The free receptions from the first transmission should be transferred to these wake-ups. The first wake-up is then handled as a non-interfering transmission with no free receptions and the second is handled as an overlapping transmission with the free receptions from both transmissions.

Since the local data captured by the sensing device runs in parallel with the states of the MAC layer, the specific sensing cost cannot be separated from the communication costs. Consequently, parameters E_I, E_R, and E_T become overestimated. However, it can be shown that by underestimating the sensing cost as $E_S = 0$, we have compensated for this overestimation.

Conventional frequency-based estimates are used to compute P_I and P_R (we have no need for P_S when $E_S = 0$), and to save memory all energy and probability estimates for $k < 100$ are based on a limited Exponential Weighted Moving Average (EWMA). For $k \geq 100$, E_I, E_R and E_T are replaced by $E_x(k) = 0.99E_x(k-1) + 0.01e_x$, which allows the estimate to adapt to changes in the environment.

3.3 Selective Transmission

The selective transmission component is called by the network layer (see Fig. 1) whenever the sensor node needs to make a decision about a packet transmission. It makes use of E_T, E_I, E_R, P_I, and P_R from the mote statistics to estimate ρ in (3), which is then used in (2).

4 Experiments and Evaluation

We ran the selective transmission implementation on top of the TinyOS CC2420 LPL layer on a number of Tmote Sky platforms. All experiments average 5 similar runs with an LPL interval of $500ms$, a data rate of one packet per $2s$, and a delay to turn off the radio after packet reception of $30ms$ (which is used as a mechanism in TinyOS to allow a transmitter to send a number of packets without a receiver turning off its radio). To avoid data periods from multiple transmitters to be synchronized, the data timer is set to a random time during the second half of the interval. Importance values of messages, which should be provided by the application layer to source nodes, are assigned according to random samples of a long ramp distribution (1,2,4,8,16,32) with decreasing probabilities. Moreover, messages are sent immediately after generated or received. The initial battery of sensor nodes is set to $200mAh$, and a sensor node is considered dead once its battery depletes. In the following, we compare the implementations based on the total importance sum of messages sent throughout the sensor node lifetime.

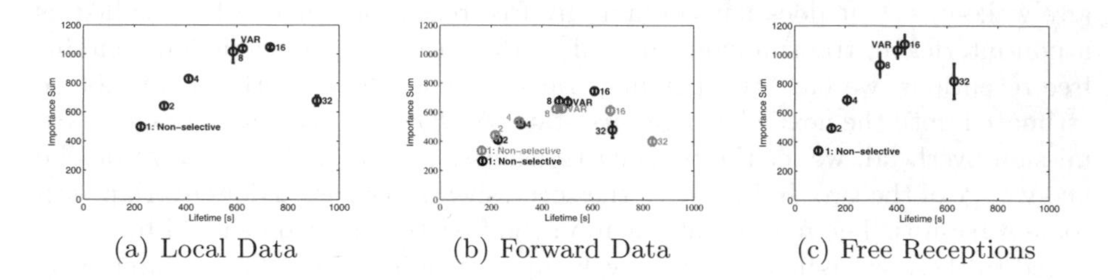

(a) Local Data (b) Forward Data (c) Free Receptions

Fig. 5. Importance sum and lifetime for variable and all fixed thresholds

To test the benefits of selective transmission based on locally sensed data we deployed two motes, a transmitter and a receiver, within radio range. The transmitter periodically senses data and makes a transmit or discard decision. Fig. 5(a) shows the average importance sum and its standard deviation, and the (average) transmitter lifetime, for the adaptive threshold (based on (2) and the energy estimates, labeled as VAR) and for all fixed thresholds (1,2,4,8,16 and 32, where the number indicates the minimum importance value that is transmitted). Selective transmission policies outperform the non-selective transmitter (threshold 1). Moreover, the adaptive threshold performance is close to the best threshold, 16. The slight differences can be explained by some suboptimal decisions during the initial steps, when the energy estimates are not accurate.

To test the selective transmission when the node receives data, we deployed three motes on a line: a non-selective transmitter, a selective forwarder, and a receiver. The transmitter periodically generates and sends packets through the forwarder (which does not sense data) to the receiver. Fig. 5(b) (black plot) shows that the importance sum of the adaptive threshold is closer to 8 than to the best constant threshold, 16. A further analysis of this case showed that the suboptimal behavior can be explained by the randomness of the importance sequence. Further tests with a periodical importance sequence (still keeping the same frequencies of each importance values), which reduce randomness, demonstrate that the variable threshold behaves optimally (Fig. 5(b), gray plot).

To test selective transmission with free receptions we placed three motes on a line again, but this time with both the transmitter and the forwarder periodically generating data and sending them to the receiver. Free receptions may happen at the forwarder when it starts to transmit local generated packets before it receives a packet from the transmitter. Fig. 5(c) shows that the adaptive threshold accurately predicts the best constant threshold in this case as well.

Fig. 6 (left) shows the final probability estimates P_I and P_R for all thresholds in the free reception scenario. The lower the threshold is, the higher the probability of reception is, because the node is less selective and therefore, P_I decreases. Fig. 6 (right) shows the final energy estimates for all thresholds in the free reception scenario. Although the current consumption during reception is slightly higher than that of transmission (see Sec. 3.1), E_T is much higher than E_R because of the longer time spent by nodes during transmission states. This explains the considerably superior performance of the selective transmission

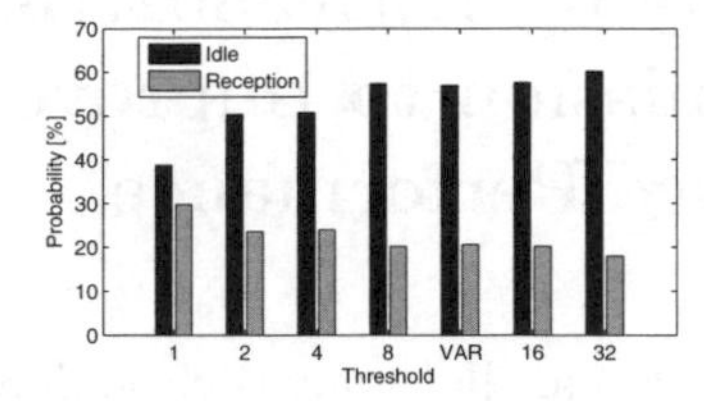

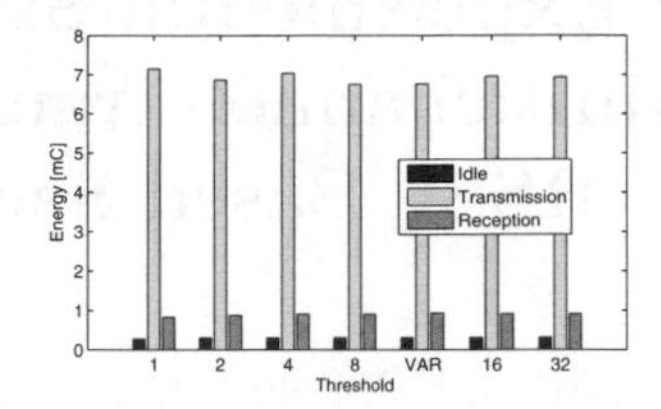

Fig. 6. Probability (left) and energy (right) estimates for all thresholds in a free reception scenario

policies with respect to a non-selective transmission in the chosen scenario. The correctness of the energy and probability estimates is implied by the fact that the adaptive threshold correctly predicts the best constant threshold. As sanity check, we can compare the energy estimates to the expected maximum transmission cost (assuming current consumption of $20mA$) of $10mC$ and see that on average it is a bit above 50% which is expected for a LPL MAC protocol.

5 Conclusion

The implementation of a selective transmission policy on top of a real LPL MAC protocol using a specific procedure to estimate energy consumption of the node states has shown how this kind of strategies can be used to extend the network lifetime and maximize the total importance sum of the transmitted messages. Future work includes the performance analysis in a larger testbed and more complex scenarios (e.g. interferences, link quality, etc).

References

1. Appadwedula, S., Veeravalli, V., Jones, D.L.: Decentralized detection with censoring sensors. IEEE Transactions on Signal Processing 56(4), 1362–1373 (2008)
2. Arroyo-Valles, R., Marques, A.G., Cid-Sueiro, J.: Optimal Selective Transmission under Energy Constraints in Sensor Networks. IEEE TMC 11(8), 1524–1538 (2009)
3. Buettner, M., Yee, G.V., Anderson, E., Han, R.: X-mac: a short preamble mac protocol for duty-cycled wireless sensor networks. In: 4th Int. Conf. on Embedded Networked Sensor Systems, SenSys 2006, New York, NY, USA, pp. 307–320 (2006)
4. Dunkels, A., Osterlind, F., Tsiftes, N., He, Z.: Software-based on-line energy estimation for sensor nodes. In: Procs. of the 4th Workshop on Embedded networked sensors, EmNets 2007, pp. 28–32. ACM, New York (2007)
5. Lei, J., Yates, R., Greenstein, L.: A generic model for optimizing single-hop transmission policy of replenishable sensors. IEEE TWC 8(2), 547–551 (2009)
6. Quek, T., Dardari, D., Win, M.: Energy efficiency of dense wireless sensor networks: To cooperate or not to cooperate. IEEE JSAC 25(2), 459–470 (2007)

An Experimental Study on IEEE 802.15.4 Multichannel Transmission to Improve RSSI–Based Service Performance

Andrea Bardella[1], Nicola Bui[1,2,3], Andrea Zanella[1], and Michele Zorzi[1,2,3]

[1] Dep. of Information Engineering, University of Padova, Italy
[2] Patavina Technologies, Padova, Italy
[3] Consorzio Ferrara Ricerche, Ferrara, Italy

Abstract. In Wireless Sensor Networks (WSNs) the majority of the devices provide access to the Received Signal Strength Indicator (RSSI), which has been used as a means to enable different services and applications like localization, geographic routing and link quality estimation. Notwithstanding the popularity of using RSSI for localization, academic research showed that RSSI-based distance estimate is rather unreliable due to the random attenuation experienced by the radio signals, as the multipath fading. In this paper we propose a simple way to improve the RSSI reliability, averaging samples collected at different frequencies by a CC2420 radio, which implements the IEEE 802.15.4 standard, both in real indoor and outdoor scenarios. For this purpose, we introduce a simple communication protocol to coordinate data exchange between nodes, that exploits multichannel transmission in order to mitigate the multipath effect that hampers ranging estimation as well as wireless communication.

1 Introduction and Related Work

Ever since the beginning of radio communication, linking the communication distance to the received signal power in a reliable way has been a hot research topic. Solving this issue would open the path for accurate localization applications [1], precise geographic routing [7], trustful self-driven robots [8] and a whole set of other context–aware systems [9].

The availability of a Received Signal Strength Indicator (RSSI) in most of commercial off-the-shelf radio transceivers has promoted the design of several RSSI-based ranging techniques that, however, suffer two major drawbacks. On the one hand, inferring the transmitter-receiver distance from the received signal strength requires a rather accurate channel propagation model [12,18]. On the other hand, the relation between distance and received signal power is very noisy due to the random attenuation phenomena that affect the radio signals, as multipath fading and shadowing [19,15].

In this paper we address these two issues in the case of radio systems based on the common IEEE 802.15.4 standard. We observe that many works focus

P.J. Marron et al. (Eds.): REALWSN 2010, LNCS 6511, pp. 154–161, 2010.

on IEEE 802.15.4 channel characteristics [18] and investigate the feasibility of using RSSI measures for ranging purposes [16]. In general, results show that RSSI-based ranging is quite poor, in particular in indoor environments [17], so that accurate localization is possible only using large number of RSSI samples [8] and/or sophisticated filtering processes to reduce the localization error [10,11].

However, to the authors' knowledge, no previous work has yet considered the possibility of exploiting the frequency diversity provided by the standard to enhance the ranging performance. In this paper, we advocate that the RSSI-based ranging accuracy can be significantly improved by considering a more accurate channel propagation model and a slightly more sophisticated communication protocol that enables the collection of RSSI samples on different frequency channels. More specifically, we first propose an Extreme Value Distribution model for the received power, which fits our empirical data better than the most common Gaussian model, both in indoor and outdoor scenarios. Second, we prove that averaging the RSSI samples collected at different carrier frequencies will mitigate the multipath fading effect, thus potentially improving the RSSI-based distance estimate at a price of a limited increase in the communication protocol complexity.

2 Channel Characterization

An extremely accurate channel model would require perfect knowledge of the environment. Clearly, such a model would lack in generality and reusability. Therefore, it is generally preferable to consider more general models that can fit a much wider set of scenarios, though with lower accuracy. A very common radio channel model that binds the received power P_{rx} to the distance d between the transmitter and the receiver is the following:

$$P_{rx}\ \mathrm{dBm} = P_{tx}\ \mathrm{dBm} + K\ \mathrm{dB} - 10\eta \log_{10}\left(\frac{d}{d_0}\right) + \Psi\,, \qquad (1)$$

where P_{tx} is the transmitted power in dBm, K is a unitless constant that depends on the enviroment, d_0 is the reference distance to be in far field conditions, η is the path loss coefficient and Ψ is a random variable that takes into account fading effects. Characterizing these parameters to the specific environment makes it possible to use the same model in different scenarios.

For instance, in a free–space environment we typically have $\eta = 2$ and K dB $= 20\log_{10}\frac{\lambda}{4\pi d_0}$, with λ the wavelength at the carrier frequency. For other common environments (in office, open space, urban and so on), K and η can be retrieved from the literature [6] or, alternatively, jointly determined minimizing the mean square error (MSE) between the model and the empirical measurements.

The characterization of the random term Ψ is, instead, more arguable. A common practice is to model Ψ as a Gaussian random variable, with zero mean and standard deviation σ_{ψ}. In this paper we advocate that, for the technology and the environments considered in this study, the model of Ψ that statistically best fits with our empirical data is the Extreme Value random variable. This model arises if we consider a received signal composed of clusters of multipath waves propagating in a non-homogeneous environment. In this case, the envelope of the received signal turns out to be Weibull distributed [13,14], with probability density function (pdf)

$$f_Z(z) = (\beta/\Omega^\beta) z^{\beta-1} e^{-(z/\Omega)^\beta},$$

where the power parameter β expresses the fading severity. In dB scale, the received signal power $P_{rx} = 10\eta \log_{10}(Z)$ turns out to have an Extreme Value distribution $f_X(x)$ with pdf

$$f_{P_{rx}}(x) = \frac{A}{\sigma} e^{(Ax-\mu)/\sigma} e^{-e^{(Ax-\mu)/\sigma}}$$

where $A = \frac{\ln 10}{10\eta}$, $\mu = \ln(\Omega)$ and $\sigma = 1/\beta$. As a result, the term Ψ in (1) is also described by an Extreme Value distribution with pdf

$$f_\Psi(x) = \beta A M e^{\beta A x} e^{-Me^{\beta A x}} = \sigma_\psi^{-1} e^{(x-\mu_\psi)/\sigma_\psi} e^{-e^{(x-\mu_\psi)/\sigma_\psi}} \tag{2}$$

with parameters $\sigma_\psi = (A\beta)^{-1}$ and $\mu_\psi = -(A\beta)^{-1} \ln M$, $M = \left(\frac{P_r^{1/\eta}}{\Omega}\right)^\beta$, and P_r denoting the mean received power (in mW).

The Maximum Likelihood estimation for d based on (1), is given by

$$\hat{d} = d_0 10^{\frac{P_{tx}+K-P_{rx}+\mu_\psi}{10\eta}} = d 10^{\frac{\Psi}{10\eta}}. \tag{3}$$

It might be worth remarking that the estimated distance $\hat{d}$ is biased. Though it is possible to correct this bias, for space constraints we do not provide any further detail on this respect. Instead, we report the relation between Ψ and the ranging error $\varepsilon_d = \hat{d} - d$, which is proportional to the distance itself d whose cumulative distribution function (cdf) turns out to be given by

$$F_{\varepsilon_d}(\alpha) = F_\Psi\left(\tilde{\psi}\right) = \begin{cases} 1 - \exp\left(-e^{(\tilde{\psi}-\mu_\psi)/\sigma_\psi}\right), & \text{if } \alpha > -d \\ 0, & \text{if } \alpha \le -d \end{cases} \tag{4}$$

where

$$\tilde{\psi} = 10\eta \log_{10}\left(1 + \frac{\alpha}{d}\right).$$

3 Multi-channel RSSI Sampling

As known, the impact of multipath on the received signal depends on the delay spread T_{rms} and the signal bandwidth B. If $T_{rms} \ll B^{-1}$ we can describe the radio propagation with a narrowband fading model, so that the received signal can be expressed as

$$r(t) = \Re\left\{u(t)e^{j2\pi f_m t}\left(\sum_n a_n(t) e^{-j\phi_n(t)}\right)\right\} \tag{5}$$

where $u(t)$ is the complex envelope of the transmitted signal, $a_n(t)$ and $\phi_n(t) = 2\pi f_m \tau_n$ are the amplitude and the phase associated with the n–th multipath component, respectively, and f_m is the carrier frequency. We observe that the phase difference between the Line of Sight (LOS) path and a reflected path is given by

$$\Delta\phi_m = 2\pi f_m \frac{\delta}{c} \tag{6}$$

where δ is the difference between the length of the two paths and c is the propagation speed of the electromagnetic wave.

Now, the IEEE 802.15.4 standard entails a transmission pulse with bandwidth of 3 MHz which can be modulated over 16 different channels, with carrier frequencies equal to[1]

$$f_m = 2405 + 5(m - 11) \text{ [MHz]}, \quad m = 11, \ldots, 26 .$$

Furthermore, typical indoor values of T_{rms}, which can be found in [5], are generally less than 100 ns, so that we can safely assume that IEEE 802.15.4 signals are affected by narrowband fading.

Nonetheless, we notice that the phase difference (6) between the direct and reflected signal components may vary significantly for sufficiently different values of m. For instance, if we consider a reflect path $\delta = 3$ m longer than the direct path, the phase difference between the two signal components that we observe in channel $m = 11$, i.e., $\Delta\phi_{11}$ differs from $\Delta\phi_{21}$ of approximately π. This suggests that the stochastic component that affects the RSSI measures may be averaged out by taking the mean value of samples collected at different frequency channels.

To sustain this claim, we designed an extremely simple communication protocol that enables the collection of RSSI samples between any pair of nodes on different channels. Basically, when a node wants to initiate the data exchange it transmits a *request* packet over the default channel (26 in our case). Such a packet carries a field with the next channel to be used for that communication. If the node receives a *reply* then the next data fragment will be sent over the new scheduled frequency. Otherwise, it assumes that the communication link is lost and returns to the default channel.

4 Experimental Campaign

This section describes the thorough experimental campaign that has been performed to collect the RSSI measurements we used to validate the channel model (1) with the Extreme Value statistical distribution for the multipath fading (2) and to sustain our claim regarding the reduction of the RSSI variations when averaging the samples collected in different channels.

For all the experiments we used Tmote Sky sensor nodes [4] mounting an isotropic antenna of known gain. These devices are equipped with the Chipcon wireless transceiver CC2420 [2] implementing the IEEE 802.15.4 standard

[1] We here respect the standard numeration of the IEEE 802.15.4 channels that conventionally goes from 11 to 26.

that specifies 16 channels with carrier frequencies $f_m = 2405 + 5(m - 11)$ MHz, $m = 11, ..., 26$. We considered three scenarios for the experimental campaign, that provide different environmental conditions. The collected data can be downloaded from the SIGNET group website [3].

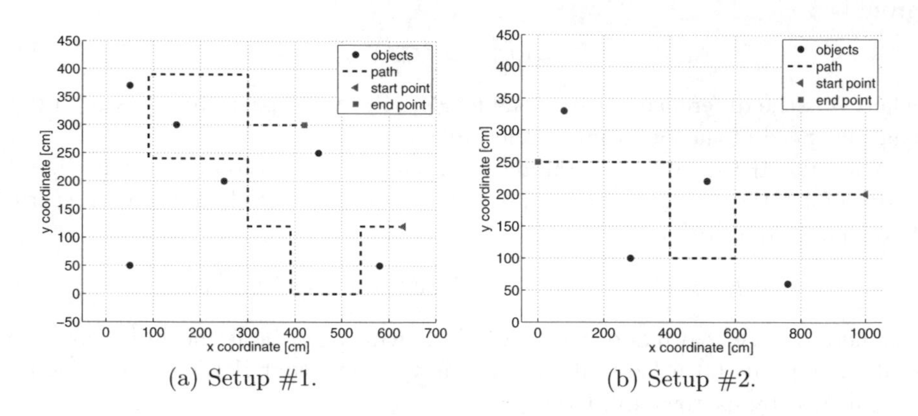

(a) Setup #1. (b) Setup #2.

Fig. 1. Paths for experimental setups

Setup #1. The sensor nodes were deployed on boxes at 50 cm from the floor. Initially we collected RSSI samples from motes deployed on a grid (24 different positions), afterwards we deployed seven nodes at known positions into the room and with another mote we moved along a pre–planned path, collecting RSSI samples in 50 locations (see Fig. 1 for node positioning in indoor setups).

Setup #2. Five sensor nodes were deployed on cones at 30 cm from the floor in an aisle and another mote was used to collect RSSI samples every 50 cm along the path. In this environment there was no furniture so that the reflections of the transmitted signal are due mainly to the floor, walls and ceiling.

Setup #3. With the same devices used for setup #2 we deployed five of them uniformly into a 15 m x 8 m area, at 80 cm from the floor, outdoor. We collected RSSI samples by each pair of nodes and then we used another node to gather some more measurements over the area from the five static nodes.

5 Results

Exploiting the collected RSSI measurements and the relative node distance information, we adopted a least mean square error criterion to obtain the channel parameters, i.e., K and η, and the standard deviation σ_ψ while we set $d_0 = 10$ cm.

With reference to setup #1 we conducted a measurement campaign to validate the channel model (1) and to evaluate the effect of multipath fading. Our experiments pointed out that it is not always necessary to estimate K and η. In fact, if the LOS condition is verified, their values are very close to those of the free–space case ($K \simeq -20$ dB and $\eta = 2$). Moreover, performing the estimation over a single channel gives us quite the same results, regardless of the particular carrier frequency, as shown in Table 1.

Table 1. Parameter estimation performed on single channels

	*ch*11	*ch*12	*ch*13	*ch*14	*ch*15	*ch*16	...	*ch*21	*ch*22	*ch*23	*ch*24	*ch*25	*ch*26
K dB	−21.7	−21.6	−21.7	−21.7	−22	−21.8	...	−21.3	−21.4	−22.1	−22	−22	−21.9
η	2.03	2.03	2	1.98	1.93	1.93	...	1.92	2.02	2.01	1.96	2	1.98
σ_ψ dB	4.8	4.4	4.5	4.4	4.3	4.5	...	4.2	4	4.1	4.3	4.4	4.2

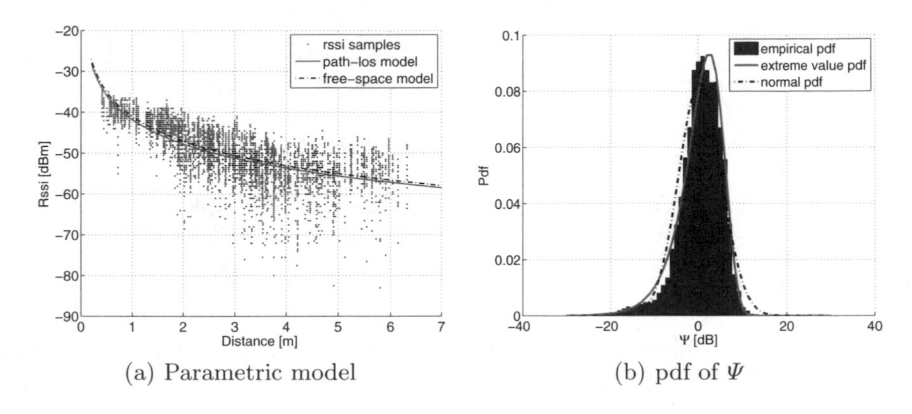

(a) Parametric model

(b) pdf of Ψ

Fig. 2. Channel models of Setup #2

Fig. 2 validates the statistical channel model of Section 2: part (a) shows the collected RSSI samples (in blue), the model in (1) with parameters estimated from the samples (solid red line) and with the free–space parameters (dash–dotted black line). In addition, part (b) verifies that Ψ has an Extreme Value distribution, hence the received power (expressed in mW) is Weibull distributed. We obtained similar results for the outdoor scenario (setup #3), though in this case we observed that the multipath fading has less impact ($\sigma_\psi = 3.5$ dB).

To evaluate the gain achieved with multichannel transmission, we collected RSSI samples from several couples of nodes, changing carrier frequency every 100 ms and sweeping all the available channels. This routine was repeated 10 times, maintaining the same experimental setup. We observed, for a particular channel and for the same link, that the RSSI samples collected at different times are quite similar, with a standard deviation less than 2 dB. Conversely, the RSSI samples collected by a pair of nodes over the 16 different channels show a high variability, with standard deviation often greater than 4 dB. Thus, as confirmed by the comparison between Fig. 3(a) and Fig. 3(b), the standard deviation of the RSSI mean reduces when samples are collected over different frequency channels in a short time period, rather than on a single channel but over a longer time interval. The experimental results, in fact, returned $K = -19.8$ dB and $\eta = 2.1$ in the two cases, whereas σ_ψ varied from 2.8 dB (frequency–average) to 4.85 (time–average), with a gain of approximately 2 dB. Instead, in outdoor environment we revealed about 1 dB of improvement. Furthermore, we observed that the same gain can be obtained by considering just four samples taken at maximum distance frequencies.

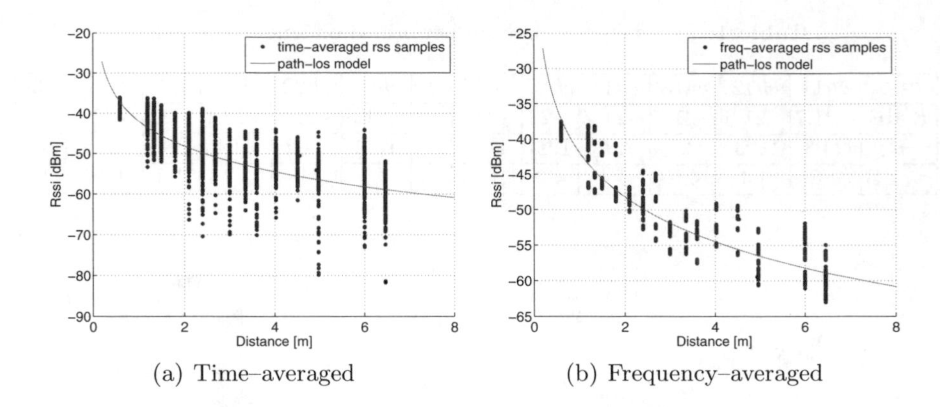

(a) Time–averaged (b) Frequency–averaged

Fig. 3. Time– and Frequency–averaged RSSI samples

6 Conclusions

In this paper, we studied the properties of the radio signal propagation in WSNs using the IEEE 802.15.4 standard. In particular, we showed that a Weibull distribution accurately fits the signal fluctuations due to multipath fading for both indoor and outdoor scenarios. We designed and developed a simple multichannel communication protocol in order to validate our analytical framework with a thorough experimental campaign. To such extent, we collected RSSI samples in different network setups, that represent typical real wireless sensor network deployments. Our results show that significant performance improvements can be obtained averaging RSSI samples over frequency.

As a final consideration, not only the channel randomness, but also physical factors such as the antennas anisotropy, the actual device sensitivity, the channel asymmetry and topology aspects impact the RSSI reliability. Thus, the network design and the device characteristics must be taken into account for proper RSSI–based service realization.

Acknowledgments

This work has been supported in part by the FP7 EU projects "SENSEI" G.A. no. 215923, http://www.ict-sensei.org, "SWAP" G.A. no. 251557, and "IoT-A" G.A. no. 257521, and by the CaRiPaRo Foundation, Italy, within the WISE-WAI project, http://cariparo.dei.unipd.it

References

1. Patwari, N., Ash, J.N., Kyperountas, S., Hero, A.O., Moses, R.L., Correal, N.S.: Locating the nodes: cooperative localization in wireless sensor networks. IEEE Signal Processing Magazine 22(4), 54–69 (2005)
2. Chipcon AS SmartRF CC2420 Datasheet. Texas Instruments Inc., (June 2004)

3. Indoor and Outdoor 802.15.4 RSSI and LQI measurements, `http://telecom.dei.unipd.it`
4. Tmote sky datasheet. MoteIv Corporation, `www.moteiv.com`
5. Indoor Propagation at 2.4 GHz, `www.wirelesscommunication.nl`
6. Goldsmith, A.: Wireless Communications. Cambridge University Press, New York (2005)
7. Zorzi, M., Rao, R.R.: Geographic random forwarding (GeRaF) for ad hoc and sensor networks: multihop performance. IEEE Transactions on Mobile Computing 2, 337–348 (2003)
8. Menegatti, E., Zanella, A., Zilli, S., Zorzi, F., Pagello, E.: Range-only SLAM with a mobile robot and a Wireless Sensor Networks. In: IEEE International Conference on Robotics and Automation (ICRA 2009) (July 2009)
9. Xueli An, R., Prasad, V., Wang, J., Niemegeers, I.G.M.M.: OPT: online person tracking system for context-awareness in wireless personal network. In: Proceedings of the 2nd International Workshop on Multi-hop ad hoc Networks: from Theory to reality, pp. 47–54 (2006)
10. Nakamura, K., Kamio, M., Watanabe, T., Kobayashi, S., Koshizuka, N., Sakamura, K.: Reliable ranging technique based on statistical RSSI analyses for an ad-hoc proximity detection system. In: IEEE International Conference on Pervasive Computing and Communications, PerCom 2009 (May 2009)
11. Chuan-Chin, P., Wan-Young, C.: Mitigation of Multipath Fading Effects to Improve Indoor RSSI Performance. IEEE Sensors Journal 8, 1884–1886 (2008)
12. Hashemi, H.: The Indoor Radio Propagation Channel. Proc. IEEE 81(7), 943–968 (1993)
13. Jacoub, M.D.: The α–μ distribution: A general fading distribution. In: Proc. IEEE Int. Symp. Personal, Indoor, Mobile Radio Communication, Lisbon, Portugal (September 2002)
14. Sagias, N.C., Karagiannidis, G.K.: Gaussian Class Multivariate Weibull Distribution: Theory and Applications in Fading Channels. IEEE Trans. on Information Theory 51(10) (October 2005)
15. Lymberopoulos, D., Lindsey, Q., Savvides, A.: An Empirical Characterization of Radio Signal Strength Variability in 3-D IEEE 802.15.4 Networks Using Monopole Antennas. ENALAB Technical Report 050501 (2005)
16. Li, X.: RSS-based location estimation with unknown pathloss model. IEEE Trans. Wireless Communications 5(12), 3626–3633 (2006)
17. Zanca, G., Zorzi, F., Zanella, A., Zorzi, M.: Experimental comparison of RSSI-based localization algorithms for indoor wireless sensor networks. In: Proceedings of the Workshop on ACM Real-world Wireless Sensor Networks (REALWSN 2008), Glasgow, Scotland, pp. 1–5 (2008)
18. Puccinelli, D., Haenggi, M.: Multipath Fading in Wireless Sensor Networks: Measurements and Interpretation. In: IWCMC 2006, Vancouver, Canada (July 2006)
19. Lindhe, M., Johansson, K.H., Bicchi, A.: An experimental study of exploiting multipath fading for robot communications. In: Proc. Robotics: Science and Systems, Atlanta, GA (2007)

Multicasting Enabled Routing Protocol Optimized for Wireless Sensor Networks

Tharindu Nanayakkara and Kasun De Zoysa

University of Colombo School of Computing, Sri Lanka
tharudn@gmail.com, kasun@ucsc.cmb.ac.lk

Abstract. TikiriMC is a wireless ad-hoc routing protocol, designed for resource constrained networking environments. It provides application programming interfaces to easily implement unicasting, broadcasting and multicasting. Flexible configuration of TikiriMC allows one to easily adopt it into a desired platform. TikiriMC uses tree network topology, where there can be many such trees in a single network. Root nodes of these multiple trees form a separate mesh network. Performance tests conclude that TikiriMC has a very low routing delay compared to other implementations.

Keywords: TikiriMC, Wireless Ad-hoc Routing, Wireless Sensor Networks, Wireless Multicast Routing.

1 Introduction

Wireless ad-hoc networking is vital on deploying Wireless Sensor Networks (WSN). Developing network protocols for WSN should be carefully designed by considering the resource constraints while providing necessary features such as multicasting. Even though there are many wireless ad-hoc routing protocols, most of them do not address the communication requirements of resource constraint WSN environments, such as low power consumption. It is a fact that, network communication is the most power consuming activity in a WSN.

It should be mentioned that, there are wireless ad-hoc routing protocols, which can be used in resource constraint environments. However there are situations where most of those protocols cannot be used because, most of them are not easily configurable to meet specific needs. For example, protocols designed for a particular hardware platform may have predefined memory and processing power limitations. The same configuration may not work with a different hardware platform even if it runs the same operating system. In addition to that, there may be application specific requirements such as memory configurations. TikiriMC is designed as a configurable ad-hoc routing protocol where a programmer can simply change some variables and create a fully customized version of it which then can be used for intended hardware platform or application.

Consequently, this research is focused on the design and development of a flexible, configurable ad-hoc routing protocol which would solve above mentioned problems while improving the efficiency of network routing.

P.J. Marron et al. (Eds.): REALWSN 2010, LNCS 6511, pp. 162–165, 2010.

2 Background

Research on wireless ad-hoc routing protocols has begun to be used with wireless devices with high computation power such as laptops and PDAs. With the dawn of wireless sensor networking these ad-hoc routing approaches have been adopted to use in low resource utilized environments. Nevertheless, as the original design was to be used with devices with high resources, most of them fails to work in a sensor networking environment. However, new routing protocols, such as Lightweight Ad-Hoc Routing Protocol [1], have been developed using the concepts and features of the existing ad-hoc routing protocols but supports low resource utilized environments.

Multicast protocols are often used to communicate with a selected subset of a large set of nodes. Existing wireless ad-hoc multicast protocols can be divided to two categories. First category forms a shared multicast tree to route packets. This approach is efficient when the nodes are static and the network topology hardly changes. Duplication of packets in the network can be reduced by using multicast trees. Adhoc Multicast Routing (AMRoute) protocol [2] and Ad hoc Multicast Routing Protocol Utilizing Increasing Id-numbers (AMRIS) [3] are examples to this category. Second category forwards multicast packets via flooding or via a mesh network. This approach is efficient when there are mobile nodes in the network. In networks with high mobility multicast trees cannot be maintained properly. Flooding ensures the packet delivery, but increases packet duplication as well. On-Demand Multicast Routing Protocol (ODMRP) [4] and Core-Assisted Mesh Protocol (CAMP) [5] are examples to this category.

3 TikiriMC Design

TikiriMC is a more efficient and effective solution for handling the unique communication requirements of resource constraint wireless sensor networks. This section includes details of the design of functionalities of TikiriMC routing protocol.

TikiriMC routing protocol has a multiple tree-based network topology. Each tree starts from its own Root node, and can span for multiple levels of descendent nodes. In a particular tree, nodes without any descendent nodes (child nodes) are called Leaf Nodes. Apart from the Root node and Leaf nodes, the rest is called Sub-Root nodes.

There can be several trees in a particular network. In such a scenario the Root nodes of those trees create a mesh network among themselves, so that inter-tree communication is possible. Intra-tree communication is handled by the Root node and relevant Sub-Root nodes of a tree. If the receiver node of a transmission is in the same tree as the sender, the packet can be routed inside the intra-tree network, if not, the root node of the sender's tree should forward data packets to the inter-tree mesh network, which will then should be received by the root node of the tree of the receiver.

TikiriMC is designed as a configurable protocol. Depending on the resource constraints of the nodes, a single tree can be configured to be varied from a single

Root node to a tree with multiple levels of descendent nodes. So as a result, the whole network topology can be changed from a forest of trees to a single tree. Furthermore, it can also be changed to a complete mesh topology.

4 Implementation and Evaluation

TikiriMC is a protocol optimized for sensor networks, so it was decided to implement it on top of the Contiki [6] real time operating system specially designed for sensor networks. Each node is implemented to run two separate processes for beaconing and controlling. Networking primitives of the Rime communication stack [7] was used to implement packet routing. Beacon process was implemented using the announcement primitive, which can be configured to broadcast a 16 bit value periodically.

We decided to do the preliminary tests of the protocol using COOJA [8] network simulator which was also a part of Contiki operating system. A node arrangement of 25 nodes were used to test the protocol and same arrangement was used in all evaluations and comparisons with other protocols.

First TikiriMC was tested for network convergence. It is a vital part of the protocol as a duly converged network can route packets more effectively and efficiently. However the network convergence was found out to be time consuming. It took 260 seconds on average to converge a network of 25 nodes.

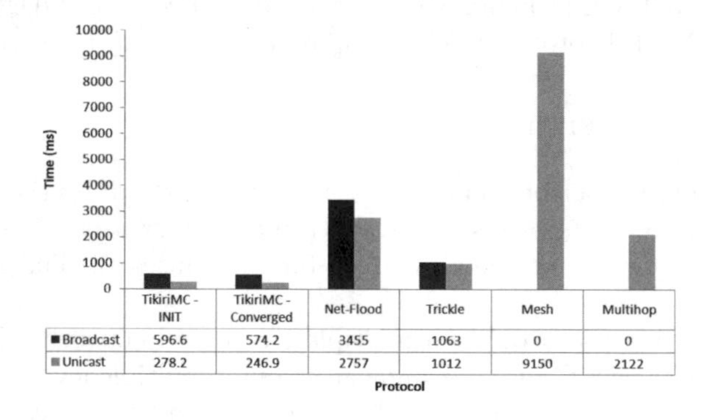

	TikiriMC - INIT	TikiriMC - Converged	Net-Flood	Trickle	Mesh	Multihop
Broadcast	596.6	574.2	3455	1063	0	0
Unicast	278.2	246.9	2757	1012	9150	2122

Fig. 1. Comparison of packet routing time of TikiriMC protocol with other protocols

Then TikiriMC was compared with four other protocols with respect to average time taken to broadcast a packet. It was tested by capturing the time taken to broadcast a 10 byte packet to all 25 nodes in the network. The results of these tests are illustrated in Fig. 1. As we can see, TikiriMC has only taken a fraction of time compared to other protocols. Nevertheless it was observed that noticeable number of duplicate packets are created in the inter-tree mesh network when sending packets. This is due to the flooding-like nature of the inter-tree mesh network.

5 Conclusions

Here, we present a new routing protocol, TikiriMC, for WSN which is capable of handling unicast, broadcast and multicast routing in resource constrained environments. This protocol uses a multiple tree topology where root of the trees form a mesh network. One interesting feature of TikiriMC is the ability to adapt it to the requirements of different hardware platforms and applications just by changing a simple configuration. TikiriMC multicasting is going to be implemented using both tree based and flooding mechanisms. This protocol is implemented on Contiki real time OS on top of Rime communication stack and preliminary tests were conducted using the COOJA network simulator. Performance evaluations convinced that the broadcasting delay of TikiriMC is very low when compared to other protocol implementations on Rime.

Acknowledgements

We appreciate the contributions by Nayanajith Laxaman (UCSC) and Kasun Hewage (UCSC). We also thank Kenneth Manjula (UCSC) for helpful comments and suggestions. We would also like to thank the anonymous reviewers for their valuable comments.

References

1. Nanayakkara, T.D., Priyadarshana, B.L., Embuldeniya, L.C., Wattegedara, R.P., Madhushanka, D.G.P., Jayawardena, C.: Lightweight ad-hoc routing protocol. In: Proceedings of the 5th SLIIT Research Symposium, PSRS 2009, Malabe, Sri Lanka, vol. 3, pp. 74–79 (December 2009)
2. Xie, J., Talpade, R.R., Mcauley, A., Liu, M.: Amroute: ad hoc multicast routing protocol. Mob. Netw. Appl. 7(6), 429–439 (2002)
3. Wu, C., Tay, Y., Toh, C.K.: Ad hoc multicast routing protocol utilizing increasing id-numbers (amris) functional specification. Internet-Draft draft-ietf-manet-amris-spec-00.txt, Internet Engineering Task Force (November 1998) work in progress
4. Lee, S.J., Gerla, M., Chiang, C.C.: On-demand multicast routing protocol. In: IEEE WCNC 1999, pp. 1298–1302 (September 1999)
5. Garcia-Luna-Aceves, J., Madruga, E.: The core-assisted mesh protocol. IEEE Journal on Selected Areas in Communications 17(8), 1380–1394 (1999)
6. Dunkels, A., Gronvall, B., Voigt, T.: Contiki - a lightweight and flexible operating system for tiny networked sensors. In: Proceedings of the 29th Annual IEEE International Conference on Local Computer Networks, LCN 2004, Washington, DC, USA, pp. 455–462 (2004)
7. Dunkels, A., Österlind, F., He, Z.: An adaptive communication architecture for wireless sensor networks. In: Proceedings of the Fifth ACM Conference on Networked Embedded Sensor Systems, SenSys 2007 (2007)
8. Eriksson, J., Österlind, F., Finne, N., Tsiftes, N., Dunkels, A., Voigt, T., Sauter, R., Marrón, P.J.: Cooja/mspsim: interoperability testing for wireless sensor networks. In: Proceedings of the 2nd International Conference on Simulation Tools and Techniques, Simutools 2009, ICST (Institute for Computer Sciences, Social-Informatics and Telecommunications Engineering), Brussels, Belgium, pp. 1–7 (2009)

GINSENG - Performance Control in Wireless Sensor Networks*

Ricardo Silva

University of Coimbra, University College Cork, University of Cyprus, Lancaster University, TUBS, SAP, SICS, GALP

Abstract. Real deployments of wireless sensor networks (WSN) are rare, and virtually all have considerable limitations when the application in critical scenarios is concerned. On one side, research in WSNs tends to favour complex and non-realistic mechanisms and protocols and, on the other side, the responsible for the critical scenarios, such as the industry, still prefer well-known but expensive analog solutions. However, the aim of the GINSENG Project is to achieve the same reliability of WSNs that the conventional analog systems provide, by controlling the network performance. In this paper we present the GINSENG architecture and the platform that have been implemented in a real scenario, considered one of the most critical in the world: an Oil Refinery.

1 Introduction

Traditionally, monitoring and control systems are analog and wired. Constituted by basic hardware and requiring complex and expensive deployments and upgrades, these systems are reliable and companies trust them. Nevertheless, wireless solutions have evolved and their low cost are making them more attractive. The idea of avoiding the deployment of thousands of cables, most of them located underground in long and inaccessible ditches, together with the amount of money that could be saved, have attracted large companies to these technologies. However, in critical scenarios, present in most industries, the only the use of reliable systems is permitted and therefore it is necessary to assure performance control of the deployed wireless systems, making them as reliable and trustworthy as the wired solutions.

In the scope of the European Project GINSENG (http://www.ict-ginseng.eu/), the consortium has been developing a tightly controlled WSN to operate in critical and unstable environments. Currently, the consortium has successfully deployed a WSN in an oil refinery in Portugal , which is used as an indicator system (sensing, no actuation) in a critical zone.

The Ginseng project focuses on controlling wireless system performance, and has targeted a set of different monitoring scenarios within the oil refinery. When

* FP7-ICT-2007-2 GINSENG: J. Sa Silva, A. Cardoso, P. Gil, J. Cecilio, P. Furtado, A. Gomes, C. Sreenan, T. O Donovan, M. Noonan, A. Klein, Z. Jerzak, U. Roedig, J. Brown, R. Eiras, J. O, L. Silva, T. Voigt, A. Dunkels, Z. He, L. Wolf, F. Bsching, W. Poettner, J. Li, V. Vassiliou, A. Pitsillides, Z. Zinonos, M. Koutroullos, C. Ioannou.

P.J. Marron et al. (Eds.): REALWSN 2010, LNCS 6511, pp. 166–169, 2010.

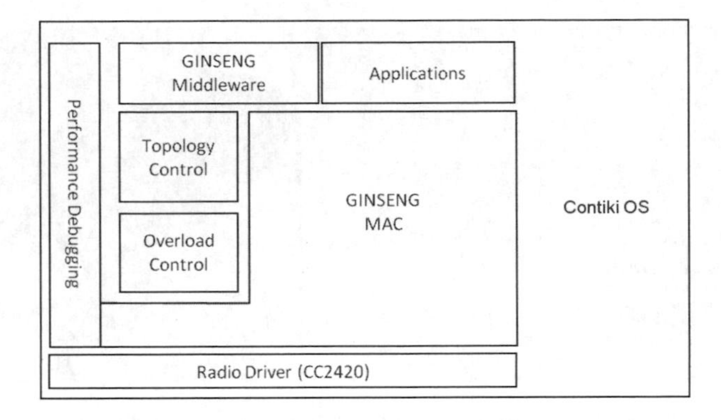

Fig. 1. GINSENG software modules

monitoring tank levels, pipes pressure, product flows or employees health , the project aims to provide a trustworthy wireless system.

To deploy a performance controlled WSN the consortium defined and implemented the architecture shown in Fig. 1.

The GINSENG architecture is based on the GINSENG MAC [1], which main function is to provide addressing and channel access control mechanisms to allow GINSENG nodes that are within radio range to communicate. It is a multi-hop system that uses an exclusive TDMA for channel access with a pre-dimensioned virtual tree topology and hierarchal addresses. The Overload Control module operates over GINSENG MAC and is responsible to drop packets that have expired or cannot be sent due to low capacity. It may also increase the priority of low priority packets, and reorder packet queues. The Topology Control is responsible for managing the tree topology, implemented by the GINSENG MAC. Performance Debugging [2] is a cross-layer module and is main function is to determine whether performance requirements are being met by the wireless sensor network. The GINSENG middleware connects the wireless sensor nodes to the high-level business applications in the backend such as ERP systems, data warehouses and advanced visualization tools. The GINSENG WSN is supported by the Contiki Operating System.

2 Real Deployment

At this stage, we have deployed a 15 nodes Wireless Sensor Network, in the refinery, in order to monitor pipe pressure and products flow in a specific critical area. From the first deployment, many lessons have been learnt. Industrial environments, such as refineries, are truly challenges for wireless communications. Besides, in critical areas, hardware components that might behave as ignition must be compliance with the ATEX directive. Therefore, each deployed mote was inserted in an ATEX compliance box and external antennas were included as inside the ATEX box standard antennas become inoperable. Fig. 2 shows the mote inside the box with the external antenna.

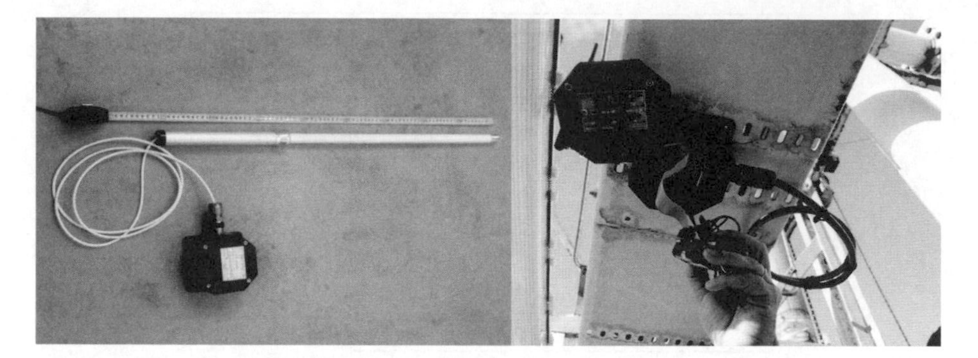

Fig. 2. TelosB motes in ATEX Boxes and 9dB Antennas

Fig. 3. Photos taken from our WSN deployed in the Petrogal oil refinery

The hardware chosen to deploy in the refinery was the TelosB mote, using the available ADC and DAC connectors to make the interface with the local pressure and flow sensors.

As previously mentioned, we used external 9dB antennas. Other models were evaluated, such as 1dB and 5dB external antennas. Scanning all available channels in different locations of the refinery, we concluded that the 9dB was the best option to assure the desired reliability. In the same study, we also concluded that channel 15 would be, in general, the best option for the location of the current network. However, any industrial environment has different patterns regarding radio communication. Reflections, refractions, absorptions, diffractions or scattering may occur in different levels, not only from place to place, but also from time to time. From our first deployment we have learnt that radio spectrum analysis is fundamental to assure a good and stable wireless communication. In Fig. 3 we can see the scenario aspect and part of the deployed network.

The deployed network is being controlled from a portable office, located near the network, in which the sink is installed. All data is received through the sink and processed locally. To analyze the system accuracy, we have compared the data received via the WSN with the data received in the Petrogal control room, via the conventional wired solution. Despite the signal noise, we realized that the sensors are quite accurate and our solution is operating as desired (Fig. 4).

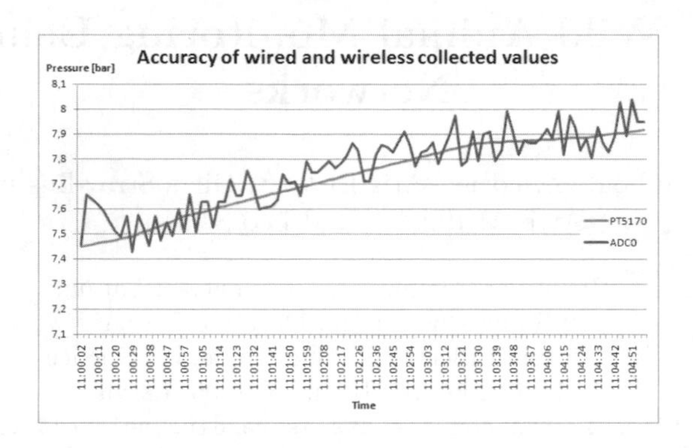

Fig. 4. Comparison of data received via the two different systems, wired (PT5170) and wireless (ADC0), from the same sensor at the same time

As mentioned before, the GINSENG middleware is responsible to provide data from the sensor network to the applications. At this stage, the middleware supports a local application responsible for locally present the real time information through an interface based on the local synoptic. The same synoptic is also provided through a web-service and therefore, remotely accessible.

Future work comprises the extension of the GINSENG network to cover other scenarios and areas in the refinery. Furthermore, all the software modules running in the motes are still under development, aiming to provide motes with more intelligence while keeping the requirements as low as possible.

References

[1] ODonovan, T., Brown, J., Roedig, U., Sreenan, C.J., Doo, J., Dunkels, A., Klein, A., Sa Silva, J., Vassiliou, V., Wolf, L.: GINSENG: Performance Control in Wireless Sensor Networks. In: Proceedings of SECON 2010 7th annual IEEE SECON Conference (2010)

[2] Pejovic, V., Sreenan, C.: PerDB: Performance Debugging for Wireless Sensor Networks. In: Proc. Of European Conference on Wireless Sensor Networks (EWSN), Poster/Demo session (2009)

LynxNet: Wild Animal Monitoring Using Sensor Networks

Reinholds Zviedris[1], Atis Elsts[1,2], Girts Strazdins[1,2], Artis Mednis[1,2], and Leo Selavo[1,2]

[1] Faculty of Computing, University of Latvia, 19 Raina Blvd., Riga, LV 1586, Latvia
[2] Institute of Electronics and Computer Science, 14 Dzerbenes Str, Riga, LV 1006, Latvia
reinholds@zviedris.com, {aelsts,gstrazdins,selavo}@acm.org, artis.mednis@edi.lv

Abstract. Monitoring wild animals, especially those that are becoming endangered (for example, lynxes and wolves) is important for biology researchers. Solutions for the monitoring already exist; however, they all have drawbacks, such as limited range or lifetime, sensing modality, reporting delays, unreliability of operation. In this work we describe our experiences in designing an improved animal monitoring sensor system and low-level software for sensor node control and communication. The target animals for this particular research are wild lynxes or canines, however it can be extended to other animal species. The LynxNet system is based on tracking collars, built around TMote Mini sensor nodes, sensors, GPS and 433MHz radio, and stationary base stations, placed at the locations that are frequented by the animals. We present preliminary field results of our radio communication range tests.

Keywords: Animal monitoring, Low power sensing, Sensor networks, Delay-tolerant networks.

1 Introduction

Monitoring the wild animal behavior and whereabouts is a challenge because the animals avoid human beings. The commercially available solutions provide monitoring devices that have limited sensing capabilities, communications requiring cellular coverage or have long data report delays [1]. We propose LynxNet system with extended sensing modality and multi hop delay tolerant communication approach. Our collaborators - biology scientists[3] aim to track Eurasian lynx (*Lynx lynx*) migration in Latvian forests. Our challenge is to achieve long-term operation with a single set of batteries in the forest environment with no energy harvesting. Our contribution includes design of simple yet persistent animal monitoring architecture for resource-constrained mobile sensor systems, development of efficient PHY and MAC layer radio communication protocols and analysis of radio communication range in field tests.

This work has been partially supported by ESF under grants Nr. 2009/0219/ 1DP/ 1.1.1.2.0/APIA/VIAA/020 and Nr. 2009/0138/1DP/1.1.2.1.2/09/IPIA/ VIAA/004.

P.J. Marron et al. (Eds.): REALWSN 2010, LNCS 6511, pp. 170–173, 2010.

2 Related Work

A number of animal monitoring sensor systems have been developed in the past few years [7][6][2][4]. The most common to our hardware is ZebraNet [7] animal tracking collar. However, not enough solar energy is available for harvesting, and lynx is smaller animal than a zebra requiring a more compact and lightweight solution.

Commercial products for GPS-based tracking are available, such as Tellus collars[1]. In comparison, LynxNet employs a wide modality of sensors in addition to GPS location, that also provide data about the surrounding environment and help to detect patterns of activities of the animal.

3 System Architecture

LynxNet is mobile sensing and sparse radio connectivity network (see Figure 1). The architecture offers animal-centric paradigm for sensing at the edge of the Internet using an opportunistic sensor networking approach.

LynxNet nodes are producing two types of packets. First type contains GPS location and fix quality information, temperature, relative humidity and amount of ambient light. One packet is formed once every hour. Second type packet contains data from 3D accelerometer and 2D gyroscope that can be used to calculate motion vector. Every 5 minutes 5 samples of data are gathered, stored in 5 packets to help with the lynx activity classification.

3.1 Hardware

LynxNet system hardware is organized as three tiers of devices: the animal collar devices (see Figure 2), the base station devices and the client devices. All devices use TMote Mini sensor nodes with TI MSP430F1611 micro-controller. Additionally LINX TRM-433-LT 433MHz transceiver (TRM) has been chosen due to its long-range characteristics.

The collar device has a radio antenna with annular operation directional diagram, GPS receiver and sensors. The prototype uses 1100 mAh 3.7V Li-Ion battery.

The base station uses four stacked half-wave dipole collinear antennas.

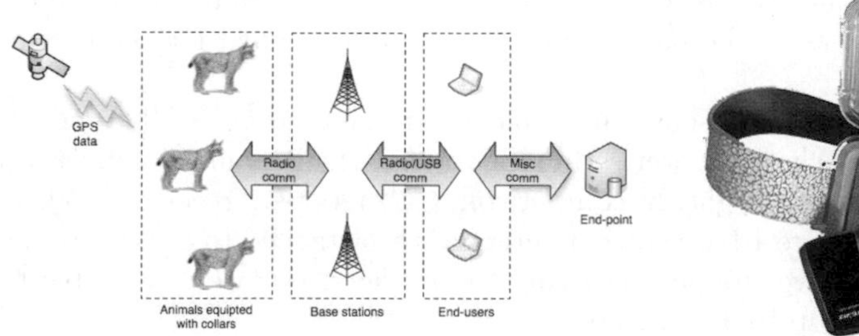

Fig. 1. LynxNet system architecture

Fig. 2. LynxNet collar device

3.2 Software

All devices were programmed using MansOS [5] operating system for networked embedded devices developed at University of Latvia. MansOS supports unix-like abstractions, programming in C and easy portability to new hardware platforms. LynxNet software extends MansOS with API for flash memory access, humidity and light sensor drivers, GPS readings parser, DCO recalibration support, the TRM chip driver, and a design of a MAC-layer communication protocol.

One of the most labor-intensive parts in the project was the development of the transceiver chip driver and PHY level communication protocol. The TRM chip has high receive sensitivity and supports symbol rates up to 10 kBdps, but provides only the simplest programming interface, and does not come with any software libraries or sample code. The chip uses OOK modulation, suitable for our project, because interference and high background noise levels are not likely in remote areas, where the system will be deployed. We implemented a packet encoding and decoding algorithm based on Manchester encoding (ME). Using ME decreases the maximal date rate to 5000 bps, but allows to decode frames with fewer errors.

On top of the physical level communication, a design for MAC protocol is build. We are going to use a CSMA-based MAC – this method is proven to work well in sparse networks where collision probability is small. The MAC protocol works differently on base stations, which are continuously listening, and mobile nodes, which have low duty cycle. All communication is initiated by the mobile nodes, which periodically poll for nearby base stations.

4 Evaluation

The LynxNet hardware field tests were performed in two locations: Rumbula airfield and Sampeteris forest, Latvia. A dog was used instead of a lynx for the better control. The animal, equipped with LynxNet collar device, was moving away from the base station. The amount of received packets was measured after every 50 meters. The collar device was sending out 22 byte packets as in real deployment situation using encapsulation described in Section 3.2.

First, we tested the hardware with TRM radio and then for comparison with CC2420 radio. Once the TRM reception tests started to fail due to the distance, we attached headphones to the base station device and listened for the carrier (beep) signal.

The results of the tests are shown in Table 1. At the distance of 300 meters we received no more packets but were still able to hear the beeping sound for up to 500 meters in the line of sight. In comparison, CC2420 tests received packets up to 165 meters. The forest base station antenna was attached to a tree. At the distance of 300 meters we stopped receiving the packets, but were still able to hear the carrier signal up to 350 meters.

TRM tests showed, that our collar antenna has a significant directional behavior. The best reception was achieved when the animal was standing with a

Table 1. Radio field test results

Distance (m)	% of packets received		RSSI (max = 4095)	
	Airfield	Forest	Airfield	Forest
50	80-100%	80-100%	2700	2800-3000
100	80%	80-100%	2200	2200-2500
150	80%	70-90%	1900-2000	2300
200	40-80%	10-50%	1600-1800	1600-1800
250	10-80%	20-50%	1600-1800	1600-1800

Table 2. Collar energy consumption

Mode	Active		mW
	sec/hour	mW	daily
Sleep	3527.123	0.033	0.78
GPS	60.0	218.79	87.52
Sensors	12.0	37.62	3.01
Radio RX	0.246	31.68	0.05
Radio TX	0.631	52.14	0.22
		Total:	91.57

side towards the base station. As expected, at 433MHz frequency the radio was less affected by the obstacles than at 2.4GHz.

We also measured energy consumption by the collar device and estimated savings with respect to the duty cycle as seen in Table 2. Based on this information and the current configuration of the prototype the lifetime is 1.5 months. Changing the GPS to MN5010HS and using a battery with greater capacity such as Enix Energies 800040 of 6800mAh 3.75V the lifetime would be extended up to 15 months.

5 Conclusion and Future Work

In this paper, we have presented our experiences designing LynxNet – an animal monitoring system in the wild. We have created a hardware prototype of a highly mobile, energy-efficient monitoring system that gathers accurate GPS position and multimodal sensor data and disseminates it through the system of delay tolerant network nodes to the consumer. Our field tests show that radio communication range of 200-250m is achievable and should be considered in further system design.

The future work includes further evaluation of collar device after longer deployments, selection of optimal components and robust packaging design.

References

1. Followit Wildlife, http://www.followit.se/wildlife/
2. GPS GSM lynx tracking in the Bavarian Forest National Park, http://www.environmental-studies.de/projects/24/GPS-lynx-tracking/gps-lynx-tracking.html
3. Latvian State Forest Research Institute Silava, http://www.silava.lv
4. POST - Pacific Ocean Shelf Tracking Project, http://postcoml.org/
5. Strazdins, G., Elsts, A., Selavo, L.: MansOS: Easy to Use, Portable and Resource Efficient Operating System for Networked Embedded Devices. In: Proc. SenSys 2010 (2010)
6. Wark, T., Crossman, C., Hu, W.,et al.: The design and evaluation of a mobile sensor/actuator network for autonomous animal control. In: Proc. IPSN 2007, pp. 206–215 (2007)
7. Zhang, P., Sadler, C., Lyon, S., Martonosi, M.: Hardware design experiences in ZebraNet. In: Proc. SenSys 2004, pp. 227–238 (2004)

Demo Abstract: Bridging the Gap between Simulated Sensor Nodes and the Real World

Tobias Baumgartner[1], Daniel Bimschas[2], Sándor Fekete[1], Stefan Fischer[2], Alexander Kröller[1], Max Pagel[1], and Dennis Pfisterer[2]

[1] Braunschweig Institute of Technology, IBR, Algorithms Group, Germany
[2] Institute of Telematics, University of Lübeck, Germany
{t.baumgartner,s.fekete,a.kroeller,m.pagel}@tu-bs.de,
{bimschas,fischer,pfisterer}@itm.uni-luebeck.de

Abstract. We present an architecture for the interconnection of simulated sensor nodes and real node hardware. The simulator is therefore running in real-time, and the simulated nodes are able to exchange messages with real sensor nodes as if they were sent over the radio. This runs fully transparent for the application—and is well suitable for debugging purposes and general algorithm development. It is even possible to use exactly the same algorithm implementation for both simulated nodes and real sensors.

Keywords: Sensor Networks, Simulation, Testbeds, Virtual Links.

1 Introduction

Algorithm development for wireless sensor networks (WSNs) is still a challenging task. It involves embedded programming on tiny micro-controllers with well-known problems such as alignment issues, unpredictability of interrupt service routines (ISRs), and a general lack of debugging possibilities. Furthermore, algorithms are mostly distributed, and thus potential errors may only occur in specific situations and are unreproducible due to oscillator variances on the nodes or message loss. A common approach is to run algorithms in simulators before testing on a real testbed. However, results from simulation are often not comparable with real-world experiments, especially when different implementations are used for the simulated nodes and real sensor nodes.

Different solutions have been presented over the past years to obviate these problems. With TOSSIM [5], it is possible to run the same code in a simulator and on real nodes. A similar approach has been presented by Wittenburg and Schiller [9], who extended the ns-2 simulator to run ScatterWeb applications—again, without changing any line of application code. With COOJA [6], it is possible to run Contiki applications before flashing them onto a sensor node. The common denominator of these approaches is that code is either run in a simulator or on hardware—without having a link between simulation and experiment, and thus without appropriate debugging possibilities when an error occurs in the testbed. To overcome these drawbacks Österlind et al. [7] presented an approach

P.J. Marron et al. (Eds.): REALWSN 2010, LNCS 6511, pp. 174–177, 2010.

with sensor network checkpointing. The state of all nodes in a testbed can be saved, and put into a simulator for further debugging. In addition, the nodes' states can also be transferred from the simulator back to the testbed.

In our demonstration, we go one step further. We enable simulated nodes to directly communicate with real sensor nodes, whereby it is possible to run the same application code in the simulator and on the nodes. The architecture is based on so called virtual links [1,3], which enable the connection between two nodes that are not in direct communication range. Messages that are sent by a node are automatically passed to a gateway, which in turn injects the message in the simulator, where it is received by a simulated node. The other way around, messages can also be sent from the simulator to a real node over the same connection.

The technique allows for embedding real sensor nodes into arbitrary topologies. Nodes can be placed at critical sections in the network, e.g. to evaluate the behavior of real nodes when being the bottleneck of a complex algorithm that is run in a large-scale simulation.

In Section 2, the architecture of the system, which allows to connect simulated nodes to real sensor nodes is presented. Section 3 describes the mode of operation of our demonstration.

2 System Architecture

The overall architecture has been developed in the context of the EU-project WISEBED [8], which aims at the interconnection of multiple testbeds scattered around Europe. These interconnected testbeds can be configured to behave like a single testbed, hiding the actual distribution to the sensor node application. This is achieved by establishing the aforementioned virtual links between individual nodes. In addition to linking real testbeds, a simulator can be integrated into the testbed—with virtual links between simulated and real nodes, see Fig. 1(a).

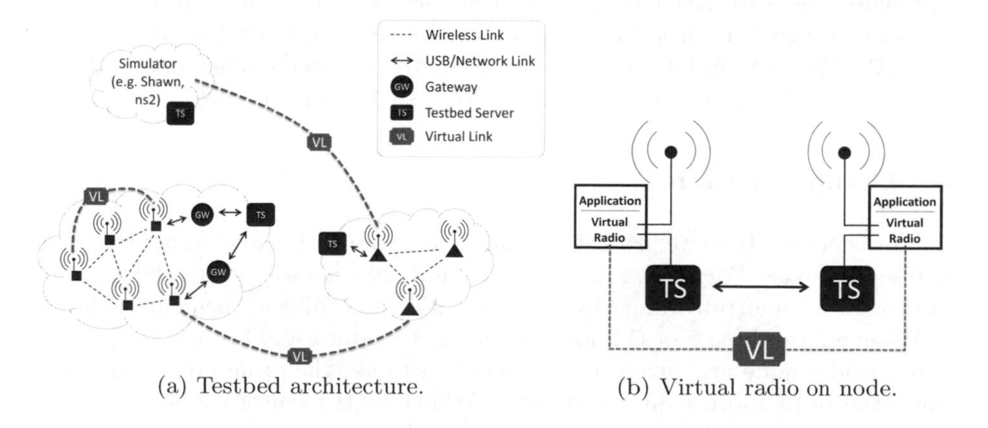

(a) Testbed architecture. (b) Virtual radio on node.

Fig. 1. Wisebed Architecture for Virtual Links

Application code can be developed using the Wiselib [2], a platform independent C++ algorithm library for heterogeneous sensor networks. Applications using the Wiselib can be compiled for several platforms such as iSense, Contiki, or even simulators. Furthermore, the Wiselib comes with built-in support for virtual links.

Virtual Links. With virtual links, one can connect different sensor node testbeds, appearing to the user as only one large testbed. Nodes in the different testbeds are linked to each other, behaving as if they were in direct communication range. A virtual link transparently tunnels messages between these nodes through the Internet, injecting the message at the destination node as if it was received via the radio. As shown in Fig. 1(b), we use a virtual radio that in turn uses both the real hardware radio and a serial connection to a gateway to be able to send/receive either via radio or a gateway. This happens transparently to the application, depending on the current set of virtual links that are configured for the individual node.

Simulation Environment. We extended the simulator Shawn [4] to be integrated in a virtual topology. First, we added the ability for real-time simulation, so that applications are executed at the same speed as on real sensor nodes. Second, basic multi-threading capability was added to be able to inject messages from real nodes at run-time. Finally, the Wisebed Web Service API was implemented to allow the connection to a real testbed—this way, nodes are not aware whether they communicate with a simulated node or with a real one.

Testbed Access. For the interconnection of a testbed with either other testbeds or a simulator, we developed a Java-based gateway software, which connects itself to the nodes inside the testbed and exposes them via the Wisebed Web Service APIs to the outside. These APIs provide a set of operations that allow researchers to manage the testbed nodes and run experiments. These include but are not limited to re-programming the nodes, collecting debug messages and sending commands to them. In addition, we added virtual link support, so that messages can be received via the Web Service API, or sent to other testbed instances via the Internet. In order to allow higher message rates than possible using the Web Service interface, we furthermore integrated a much more efficient message-based interface that uses direct TCP socket connections.

3 Demonstration

In our demonstration, we present a connection between the Shawn simulator and real sensor nodes. We use the visualization module of Shawn, where nodes can be drawn at run-time, providing a live representation according to their current state.

We have also three types of nodes: Sensors, actuators, and bridges. The sensor and a bridge node are linked to one Shawn instance, the same bridge node and the actuator to another Shawn instance. Whenever the sensor—a light sensor—detects an event (darkness or lightness), it sends a message to the actuator. The message is thereby routed through the first Shawn instance to the bridge node, which passes it to the second Shawn instance. From there, it is routed to the actuator—a light, which is turned on or off.

Acknowledgement. This work has been partially supported by the European Union under contract number ICT-2008-224460 (WISEBED).

References

1. Baumgartner, T., Chatzigiannakis, I., Danckwardt, M., Koninis, C., Kröller, A., Mylonas, G., Pfisterer, D., Porter, B.: In: Silva, J.S., Krishnamachari, B., Boavida, F.L. (eds.) EWSN 2010 LNCS, vol. 5970, pp. 210–223. Springer, Heidelberg (2010)
2. Baumgartner, T., Chatzigiannakis, I., Fekete, S.P., Koninis, C., Kröller, A., Pyrgelis, A.: Wiselib: A generic algorithm library for heterogeneous sensor networks. In: Silva, J.S., Krishnamachari, B., Boavida, F. L.(eds.) EWSN 2010 LNCS, vol. 5970, pp. 162–177. Springer, Heidelberg (2010)
3. Bimschas, D., Danckwardt, M., Pfisterer, D., Fischer, S., Baumgartner, T., Fekete, S.P., Kröller, A.: Topology virtualization for wireless sensor network testbeds. In: Proceedings of the 6th International ICST Conference on Testbeds and Research Infrastructures for the Development of Networks and Communities (TridentCom 2010), Berlin, Germany. ICST (May 2010)
4. Kröller, A., Pfisterer, D., Buschmann, C., Fekete, S.P., Fischer, S.: Shawn: A new approach to simulating wireless sensor networks. In: Proceedings of the 3rd Symposium on Design, Analysis, and Simulation of Distributed Systems (DASD 2005), pp. 117–124 (2005)
5. Levis, P., Lee, N., Welsh, M., Culler, D.: Tossim: accurate and scalable simulation of entire tinyos applications. In: Proceedings of the 1st International Conference on Embedded Networked Sensor Systems, SenSys 2003, pp. 126–137. ACM, New York (2003)
6. Österlind, F., Dunkels, A., Eriksson, J., Finne, N., Voigt, T.: Cross-level sensor network simulation with cooja. In: Proceedings of the First IEEE International Workshop on Practical Issues in Building Sensor Network Applications (SenseApp 2006), Tampa, Florida, USA (November 2006)
7. Österlind, F., Dunkels, A., Voigt, T., Tsiftes, N., Eriksson, J., Finne, N.: Sensornet checkpointing: Enabling repeatability in testbeds and realism in simulations. In: Roedig, U., Sreenan, C.J. (eds.) EWSN 2009. LNCS, vol. 5432, pp. 343–357. Springer, Heidelberg (2009)
8. Seventh Framework Programme FP7 - Information and Communication Technologies. Wireless Sensor Networks Testbed Project (WISEBED), ongoing project since (June 2008), `http://www.wisebed.eu`
9. Wittenburg, G., Schiller, J.: Running real-world software on simulated wireless sensor nodes. In: Proceedings of the ACM Workshop on Real-World Wireless Sensor Networks (REALWSN 2006), Uppsala, Sweden, pp. 7–11 (June 2006)

A Mote-in-the-Loop Approach for Exploring Communication Strategies for Sensor Networks

Minyan Hong[1], Erik Björnemo[1], and Thiemo Voigt[2]

[1] Uppsala University, Sweden
[2] Swedish Institute of Computer Science (SICS), Kista, Sweden

1 Introduction

Sensor networks are being deployed in a range of different environments, such as industry plants, rainforests and offices. Each environment has its own characteristics and the appropriate communication strategy will differ accordingly – packet sizes, retransmission schemes, error correcting codes, etc. It is, however, difficult to investigate the most appropriate communication strategies for the environment of an intended deployment. On the one hand, simulations are seldom realistic enough as they do not model the environment in every intricate detail. On the other hand, real-world experiments with deployed nodes are important but time-consuming, difficult to repeat, and to some extent dependent on hardware and software. For example, a bug in the software might make measurements collected during an extensive time useless. We need an easier way of testing which still captures realistic communication environments and provides repeatability.

We propose a new approach to investigate communication strategies. Our approach uses a combination of on-site radio channel and interference measurements, real sensor network hardware as well as a signal analyser and a signal generator. The advantage of our approach is that once the channel measurements are made, we have a deterministic and repeatable way of investigating the most suitable communication strategy in the lab and for different sensor node hardware. Additionally, we can quickly test new hardware and new implementations by simply recording new packets.

2 Approach

The setup consists of two motes, a vector signal analyser (VSA) and a vector signal generator (VSG)[1], see Figure 2.

A modern vector signal analyser/generator is an advanced instrument with the following typical characteristics: Large frequency range; Large signal bandwidth; Accurate power reading/setting. These features render the instrument flexible and facilitate tests outside the reach of mote-to-mote communication such as the transmission of recorded signals at very precisely set power levels.

The motes are TmoteSky sensor nodes [3] which feature a CC2420 radio and run the Contiki operating system. The sending mote sends one or more packets that the

[1] In our case the 2810 VSA and the 2910 VSG from Keithley.

P.J. Marron et al. (Eds.): REALWSN 2010, LNCS 6511, pp. 178–181, 2010.

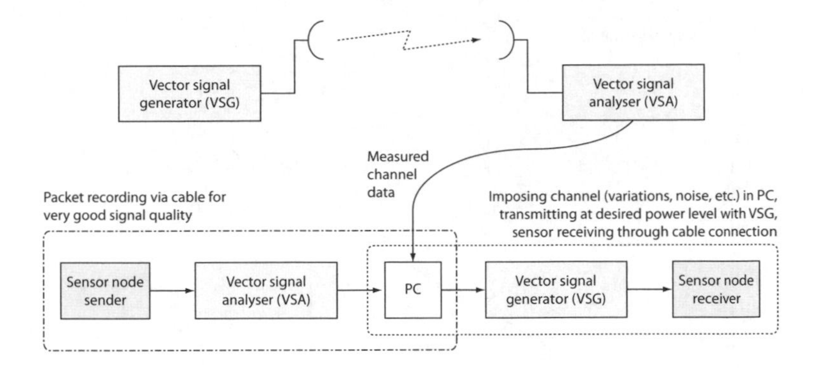

Fig. 1. Experimental setup for repeatable testing of communication strategies using real channel data

signal analyser resolves to in-phase and quadrature (IQ) values that can be stored on the PC, for example, using Matlab. The radio communication between mote and signal analyser is via a cable to avoid external interference and achieve large signal-to-noise ratio (> 60 dB). With software on the PC, we can instruct the signal generator to replay the packets received by the signal analyser and transmit them to the mote as depicted in Figure 2. We can further vary the output power of the signal generator, e.g. according to measured channel gains. Moreover, we can modify the outgoing signals by e.g. adding measured or simulated interference and noise. In particular, we are able to collect measured channel data from different environments to emulate the impact of the environment on communation. This way, we expect to be able to find communcation strategies tailored to the environment. At the receiving mote we can measure e.g. packet reception rate but also retrieve the received signal strength indicator (RSSI), the link quality indicator (LQI) and noise floor values from the on-board radio.

3 Evaluation and Proof of Concept

3.1 Basic RSSI Experiment

We verify the CC2420's RSSI readings by repeatedly replaying a recorded packet at increasing power levels. Figure 2 shows the results over the power range in which the motes actually receives the packets and can hence measure and report the RSSI (down to approximately -95 dBm). The figure shows the expected overall linear relationship, with a small variance in the RSSI readings. However, we specifically note two regions – at output powers of -40 dBm and -25 dBm – where the linear relationship between RSSI and VSG output power is disturbed and the sample variance is larger. This reflects an inaccuracy of in the RSSI reading mechanism that also Chen and Terzis have observed [2]. Note that this inaccuracy thus *confirms* the correctness of our approach.

3.2 Repeatable Test of Communication in Fading Channels

While real world deployments in one respect constitute the ultimate test of a sensor network and its communication strategies, it can be very difficult to compare results for

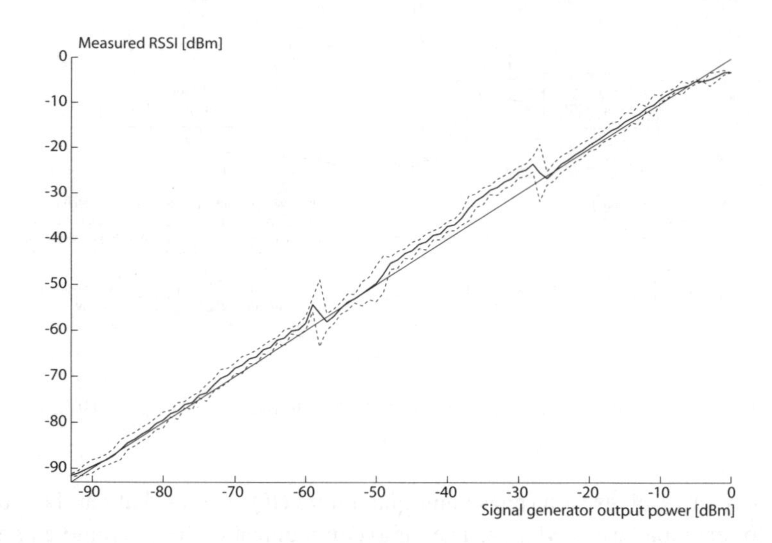

Fig. 2. RSSI from the CC2420 as a function of the VSG signal output power. The mean $\pm$ 3 standard deviations are given together with the ideal straight-line response. The curves are based on 10000 RSSI readings per power setting.

different deployments at different times. One reason is that variations in link quality – channel fading – are different at different times and locations. During research and development, it is therefore desirable to have a repeatable approach which still is much more realistic than simulation. Our proposed approach is a step in this direction, and we here show an example of the results we can achieve using real-world channel data. The channel characteristics used here consists of traces that we have collected in office and forest environments [1].

In Figure 3 the general procedure for including fading and interference is depicted. Note that there is a choice when it comes to the thermal noise as it can either be introduced artificially in the PC, as part of $n(\tau, t)$, or by using the real receiver noise and a scaled output power from the VSG. We used the latter approach and studied only the channel impact without interference for illustrative purposes. By the use of 10000 packets for each received average signal-to-noise ratio, we obtained packet error rate curves for three cases: No fading, measured office fading and measured forest fading. The channel data was applied so that block fading was achieved, that is a fairly constant channel during packet transmissions.

The results in Figure 4 show how the fading introduces error floors, starting at packet error rates of around 3 percent. The difference between the fading channels is not as extreme as one might expect, but it should be noted that the terms "line of sight" and "non line of sight" are inadequate to describe the difference. In fact, the office setting allowed some penetration through walls which resulted in "partial line of sight" (non-Rayleigh fading). Additionally, the forest setting was not pure line of sight because of the antennas being very close to the ground.

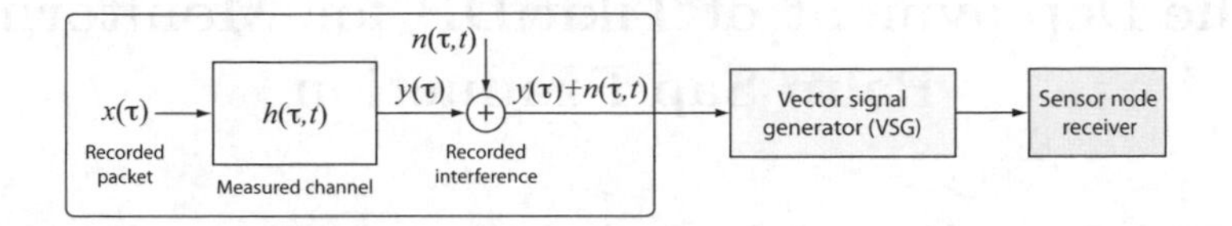

Fig. 3. Introduction of fading channel and interference. The packet $x(\tau)$, the channel $h(\tau, t)$ and the interference $n(\tau, t)$ are all complex-valued to contain both amplitude and phase information. The variable t shows that the channel and the interference can have time-varying characteristics.

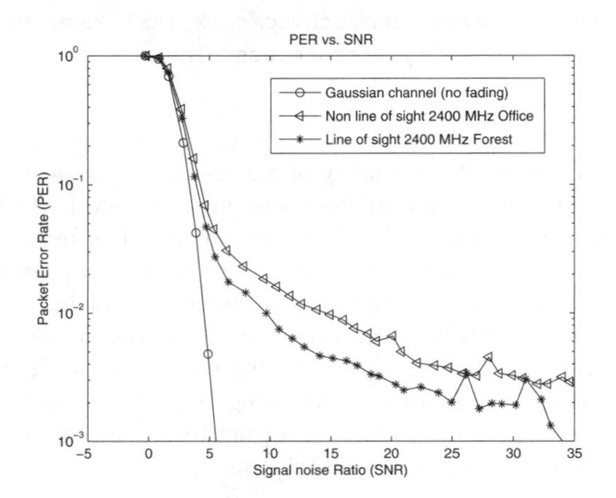

Fig. 4. Packet error rate (PER) for a range of signal-to-noise ratios (SNR). The channel was block fading, that is to say roughly constant during a packet but changing on an inter-packet time scale.

4 Conclusions

We have presented Mote-in-the-Loop, a new approach for communication strategy exploration. With some further extensions such as feedback from the sensor node to the VSG in order to trigger retransmission we believe that Mote-in-the-Loop will become a powerful and useful tool.

Acknowledgements. This work was funded by the Uppsala VINN Excellence Center for Wireless Sensor Networks WISENET, partly funded by VINNOVA.

References

1. Björnemo, E.: Energy Constrained Wireless Sensor Networks: Communication Principles and Sensing Aspects. Institutionen för teknikvetenskaper, Uppsala University (2009)
2. Chen, Y., Terzis, A.: On the Mechanisms and Effects of Calibrating RSSI Measurements for 802.15. 4 Radios. In: European Conference on Wireless Sensor Networks, Coimbra, Portugal (February 2010)
3. Polastre, J., Szewczyk, R., Culler, D.: Telos: Enabling ultra-low power wireless research. In: Proc. IPSN/SPOTS 2005, Los Angeles, CA, USA (April 2005)

The Deployment of TikiriDB for Monitoring Palm Sap Production

Asanka P. Sayakkara, W.S.N. Prabath Senanayake, Kasun Hewage, Nayanajith M. Laxaman, and Kasun De Zoysa

University of Colombo School of Computing,
No. 35, Reid avenue, Colombo 7, Sri Lanka
{asanka.code,prabathws}@gmail.com, kch@ucsc.lk, {nml,kasun}@ucsc.cmb.ac.lk
http://score.ucsc.lk

Abstract. Nowadays, the industry of harvesting palm sap in Sri Lanka is facing many problems due to theft and environmental affects. In this paper, we propose a solution for the particular problem by using a wireless sensor network based system to monitor the palm sap production of a large plantation area. We are using an enhanced version of TikiriDB which provides a database abstraction on the sensor network to make the process of data collecting and analyzing more efficient. TikiriDB can be used to reduce restrictions on accessing traditional sensor networks by enabling the user to access the sensor network using multiple queries through multiple access points simultaneously.

Keywords: Wireless Sensor Networks, Database Abstraction, Liquid level Sensor , Multiple User Access, Palm Sap.

1 Introduction

Palm sap production is a traditional industry in Sri Lanka which has a huge demand in the local market. It is obtained from different types of locally available palm trees. Owners of large palm tree estates employ rural people for tapping palm trees to collect the essence of palm flowers. The tappers climb to tree tops to slice palm tree flower and hang a special container to collect the liquid. They have to climb again to collect the containers with the liquid. Some times they use small bowser to transport the collected palm sap to main storage. Palm plantations may span over many acres of land area.

Therefore it is challenging to manage such plantations. Owners are facing the the difficulty of protecting the product from theft. Not only outsiders, but also employed tappers may involve in palm sap theft. It's practically very difficult to monitor the production by employing watchers due to the size of the land area. It's even hard to track the thefts done by the tappers who are employed in a particular palm state. In addition to that there are several more requirements to fulfill when deploying a system to solve above mentioned problem. Productivity of palm trees may depend on different environmental conditions such as temperature and humidity. Therefore, monitoring such conditions is necessary

P.J. Marron et al. (Eds.): REALWSN 2010, LNCS 6511, pp. 182–185, 2010.

to arrange necessary measures to increase productivity. This requires considerable amount of environmental data with relate to a particular palm tree and tree's productivity measures. Therefore the people in this industry, specially the owners of palm states, are seeking for a feasible and effective solution to monitor and protect the palm sap production.

2 Our Approach

We propose a solution to the mentioned problem by deploying a sensor network where each palm tree have sensors. A palm sap liquid level measuring sensor is fixed to each and every sap collecting container. Thereby, we can track the changes of palm sap liquid level in regular time intervals. Sensors to measure the environmental conditions such as humidity and temperature are also fixed to the sap containers in each palm tree. However, to analyze and produce required reports regarding the collected measures, it is required to collect these information into a central location. Since, the palm tree plantations can be spread over many acres of land area, its not effective and feasible to use wired communication. Hence, using a wireless sensor network would be an effective way of solving the problem.

Wireless sensor networks (WSN) have emerged a new wave in the community of researchers of various fields such as computer science, health care, habitat monitoring, military and disaster management etc. Since most of the devices used for WSN applications are so small, researchers have been able to obtain information from environments where they couldn't reach before. These scare resourced devices are networked through special protocols to communicate with similar motes in the network and with external computers. Therefore, using WSN requires a considerable amount of technical knowledge about the devices, protocols used, sensors, etc. However, most of the users of wireless sensor networks are not technical people i.e. zoologist, military personal, medical doctors, etc. Therefore it is better to provide them with an abstraction to the sensor network which would hide technical details from the user and at the same time provide facilities to interact with the WSN in an easier and efficient way. For example a WSN can be viewed as a file system or a database. Since, there are no WSN literal personal in the plantation, using such abstraction would be an added advantage.

File system abstractions for WSN use read and write operations to communicate with WSN where database abstractions use specific Structured Quarry Languages(SQL). Researchers from University of California, Berkeley have developed a solution called tinyDB [1] which gives a database abstraction to sensor networks using TinyOS. A similar approach has been followed by the researchers at University of Colombo School of Computing in Sri Lanka called tikiriDB [2] which also gives a database abstraction layer to sensor networks using Contiki OS. We opted to use database abstraction technologies because, palm plantations have different kinds of people who have different interests regarding the information related to the palm plantation, and database abstraction gives a flexible way to develop applications to meet the vague requirements.

For example, a botanist in the plantation may require sensor readings in an increased sample rate to measure the correlation between environmental factors and palm sap productivity for a particular tree. One solution would be to set sensors to take readings at the maximum rate they can perform. This would increase the battery consumption of sensors and hence would wear out the batteries very quickly. However, if we use a database abstraction to WSN, it is possible to specify what data required in which rate in the query. This would save a considerable amount of power with regard to previous situation. If a file system abstraction is used, implementation would be more complicated than issuing a query.

It is a requirement that the solution provides access to the WSN at any location for different parties. For an example when a bowser collects the liquid from the containers which were hanged on palm trees, wireless node in the bowser should be able to gather the information about the amount of liquid in each container of palm trees by sending a query to each tree. Therefore, the solution should be able to handle multiple concurrent access. TinyDB, doesnt allow a user to query the WSN by accessing it from any sensor node. It is only allowed to access though a defined root node for a particular WSN. However, TikiriDB database abstraction supports shared WSN system with multiple access points. Therefore, we opted tikiriDB as the database abstraction when providing the solution for palm tree plantation.

3 Implementation

We have done several enhancements to TikiriDB to adapt it into our solution such as functionality to create storage pointers, and enhancing querying language to handle event queries [3]. Storage points is a concept to store data temporarily in the nodes itself until that stored data are collected. Event queries let the user to set a query to be executed when a special event is triggered.

According to the figure 1, the system has a very modular architecture with fully distributed components over the WSN. All sensor nodes are embedded in

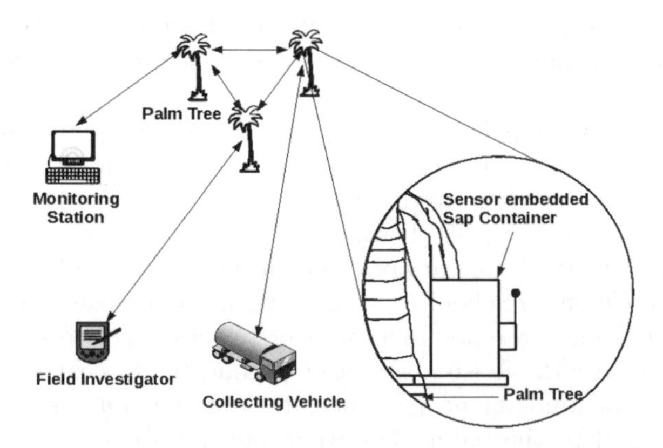

Fig. 1. High Level Overview of the System

specifically designed container which is capable of measuring the liquid level, temperature, humidity, and pressure caused due to the weight of the liquid. A combination of sensor values are used to come to a conclusion on palm theft. In addition to the automated data collection, the sensor network provides functionalities to send queries and retrieve data in real time.

At the calibration stage of the device which we are creating to get information about the collected palm sap, we identified a requirement of obtaining external variables which affects the rate of flow of palm sap. The flow of palm sap may depend on temperature, humidity, and some other factors which controls the productivity of palm sap such as the age of a particular tree. However, temperature, humidity can be obtained from the sensor nodes at trees and other factors can be obtained from external sources of data. These, data can further be used to make predictions on future productivity of a tree and total palm sap production.

4 Conclusions and Future Works

Real world deployment of WSN applications involves a lot of effort in calibrating sensors and configuring hardware components to meet the required functionality from the sensor network. Therefore the approaches to simplify the complexity by giving different abstractions can play a major role in real world deployments of WSN systems. Even though WSN based systems can show good results in experimental level, they can be inapplicable in real world implementations. Therefore the best way to evaluate a WSN system is applying it in real world problems. The initial works on palm sap production monitoring system has taught us a lot about real WSN.

By working on this palm sap production monitoring system, we realized that tikiriDB can be a part of many real world WSN deployments which has similar requirements to this project. From its original design, it has the flexibility that most of the WSN applications in the real world needs. Currently, tikiriDB developers are working on further enhancements and developments of it to provide a comprehensive database abstraction for Contiki based WSN. Therefore we can expect that it will appear in more and more real world applications in the future.

References

1. Madden, S., Franklin, M.J., Hellerstein, J.M., Hong, W.: Tinydb: An acqusitional query processing system for sensor networks (2005)
2. Laxaman, N.M., Goonatillake, M.D.J.S., Zoysa, K.D.: Tikiridb: Shared wireless sensor network database for multi-user data access (2010)
3. Madden, S., Franklin, M.J., Hellerstein, J.M.: Design of an acquisitional query processor for sensor networks (2003)

Cooperative Virtual Memory for Sensor Nodes

Torsten Teubler, Jan Pinkowski, and Horst Hellbrück

Lübeck University of Applied Sciences
{hellbrueck,teubler}@fh-luebeck.de
{pinkowsj}@stud.fh-luebeck.de
http://www.cosa.fh-luebeck.de

Abstract. Wireless sensor networks (WSN) have unique challenges and constraints. Sensor nodes e.g. have tough memory limitations. However, the latest advances in WSN research direct for an implementation of lightweight versions of Internet protocols like IPv6, TCP, and HTTP on sensor nodes. These protocols have challenging requirements. Especially, memory consumption of these protocols is often higher than the physical RAM that microcontrollers have integrated. Therefore, we suggest an approach for virtual memory providing more memory than the available RAM. As microcontrollers do not include a memory management unit the usage of memory is implemented in cooperative fashion based on the C standard library function malloc and free. We suggest an underlying file system and a hardware abstraction layer to support various external or internal memory devices like Flash or EEPROM. In this work in progress we present an API, some implementation details and preliminary results including future work.

Keywords: Wireless Sensor Networks, Virtualization, Application.

1 Introduction

The progress of research in sensor networks for the last ten years is remarkable and driven by new protocols, enhanced algorithms and more powerful applications. Todays WSN Operating Systems provide lightweight web servers and reduced functional TCP/IP stacks with IP Version 6 support [1]. When the community started with assumptions of hardware size of 2kB RAM and some 30kB of program memory today some platforms provide 10 times more memory and even more powerful CPUs. However, this is by far not enough to keep large routing tables, neighborhood lists, or allow for buffering of messages for aggregation. In our studies with IPv6 we continuously hit the memory limit problem as protocols degraded because some buffered messages that are needed later must be dropped as the memory was needed for fresh incoming messages. There are many more situations where additional memory could enhance the performance of the system significantly.

Additionally, for many applications the energy resources remain the major limiting factor whereas delay is not a critical aspect. The reason for the acceptable sensing delay is that WSNs are designed for duty cycle of nodes less than

P.J. Marron et al. (Eds.): REALWSN 2010, LNCS 6511, pp. 186–189, 2010.

10% typically 1% to guarantee a tolerable lifetime of the nodes. As a result information that crosses a large multi-hop sensor network cannot be transferred in a single active cycle resulting in a large delay.

The basic idea is that we provide larger virtual memory in our system by using RAM and unused existing Flash and EEPROM memory in parallel. In that way, an API provides access to the memory for the user in a transparent way independent from the location of the data in the RAM, in Flash or EEPROM. As the access to EEPROM or Flash is slower compared to RAM we expect an increase in access delay which we project to be tolerable as sensor nodes need to be designed delay tolerant any way as argued in the beginning of this section.

The rest of the paper is organized as follows: In Section 2 we discuss related work. Section 3 introduces our approach and discusses the implementation. The paper will conclude after presenting preliminary results in Section 4.

2 Related Work

File systems and virtual memory are considered as components in modern operating systems (OS). However, OS for resource constrained embedded devices like sensor nodes lack these functionalities. A survey of WSN OS and their characteristics can be found in [2]. We examined the most prominent operating systems like TinyOS [5], SOS [4], MANTIS [6] and Contiki [1]. None of the above OS reported support for virtual memory.

A small size WSN OS called *t-kernel* and published in [3] is most related to our approach as it also provides virtual memory. However, virtual memory in *t-kernel* is integrated into the OS and thereby hardcoded to Flash memory whereas our approach is more flexible using a mix of Flash and EEPROM on top of a file system.

3 Concept and Implementation

Our approach can be classified as cooperative virtual memory (CVM) in the sense of cooperative multitasking. CVM offers the following advantages compared to RAM only solutions: CVM provides additional memory on demand with intuitive elegant API. CVM includes a hardware abstraction layer and can compensate insufficient size of RAM. Additionally, we can expand functionality to hibernate mode in future.

However, in technical systems many improvements are not for free. In our approach we introduce some overhead in the RAM to manage the virtual memory. Additionally, the access to the virtual memory is more complex and will be slower. We will try to minimize these drawbacks by future optimizations.

The API of CVM consists of not more than four functions. A typical scenario of the usage of cooperative virtual memory is depicted in Fig. 1. We will introduce the functions in the typical order they are called by the user of CVM.

void cvm_malloc(size_t size)* returns a pointer to an allocated memory area on the heap. The size of the allocated space is passed as parameter. If there is no

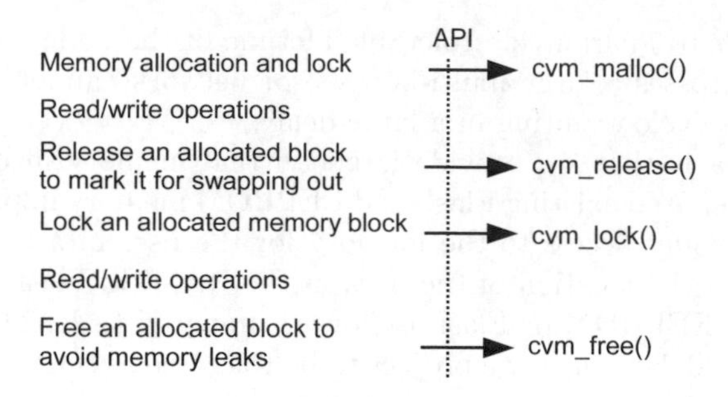

Fig. 1. Typical usage of Cooperative Virtual Memory

continuous block of memory with the specified size in RAM available, released memory blocks will be swapped out. If swap space is used completely which means no memory available, a null pointer is returned.

uint16_t cvm_release(void pBlock)* marks an allocated memory block as swappable. It returns a temporary handle for managing the swapped out data. After a call to this function the corresponding data might not be in the heap anymore and the pointers to the data are invalid.

void cvm_lock(uint16_t handle)* takes a handle as parameter. If the memory block for this handle is in the heap the pointer to the data is returned immediately. Otherwise the memory block is swapped in again. Then memory is locked.

void cvm_free(uint16_t handle) frees the memory block that has been allocated with *cvm_malloc()* and has been released.

Cooperative Virtual Memory		
stdlib.h – malloc(), free()	File System with Hardware Abstraction Layer	
RAM	EEPROM	Flash

Fig. 2. Implementation Stack of Cooperative Virtual Memory

The software structure of CVM is shown in Fig. 2. CVM is based on standard C library functions *malloc()* and *free()* for allocating RAM in the heap section. We have implemented an optimized file system for EEPROM and Flash memory.

4 Preliminary Results and Summary

Due to space restrictions we present access delay measurements in Fig. 3 exclusively. Measurements have been performed on our TriSOS hardware based on Atmega Microcontroller using the worst case setup with external I2C EEPROM. Fig. 3 demonstrates that time for swapping out (writing) RAM to the EEPROM is quite large whereas swap in (reading from the EEPROM) is much faster. In WSN protocols where round trip time is up to several seconds between sender

Block size [Byte]	Time [ms]		
	free	swap out	swap in
100	15	52	7
400	18	95	19
1024	21	186	46

Fig. 3. Time Measurements for CVM Implementation and the TriSOS Sensor Node

and receiver internal access delay of 50 ms is tolerable. Internal EEPROM or Flash is expected to be much faster as I2C-Bus introduces a substantial delay.

In this ongoing work we have introduced a novel API for virtual memory for sensor nodes together with preliminary results which provides elegant handling of memory that can be flexibly distributed to secondary storage Flash and EEPROM. An important step for improvement is decreasing the access delay for the swapping operation. Therefore, we will develop and implement an asynchronous swap in and swap out based on usage prediction.

We will implement the CVM into standard Internet protocols like routing, http in order to demonstrate the feasibility of this concept. In the future we plan to extend this concept for hibernation and post failure analysis.

Acknowledgments. This work was funded by the Federal Ministry of Education & Research of the Federal Republic of Germany (Förderkennzeichen 01BK0905, GLab). The authors alone are responsible for the content of the paper.

References

1. Dunkels, A., Grönvall, B., Voigt, T.: Contiki - a lightweight and flexible operating system for tiny networked sensors. In: Proceedings of the First IEEE Workshop on Embedded Networked Sensors (Emnets-I), Tampa, Florida, USA (November 2004)
2. Dwivedi, A.K., Tiwari, M.K., Vyas, O.P.: Operating systems for tiny networked sensors: A survey. Int. Journal of Recent Trends in Engineering 1, 152–157 (2009)
3. Gu, L., Stankovic, J.A.: t-kernel: providing reliable os support to wireless sensor networks. In: Proceedings of the 4th International Conference on Embedded Networked Sensor Systems, SenSys 2006, pp. 1–14. ACM, New York (2006)
4. Han, C.-C., Kumar, R., Shea, R., Kohler, E., Srivastava, M.: Sos: A dynamic operating system for sensor networks. In: Proceedings of the Third Int. Conference on Mobile Systems, Applications, And Services (Mobisys). ACM Press, New York (2005)
5. Levis, P., Madden, S., Polastre, J., Szewczyk, R., Whitehouse, K., Woo, A., Gay, D., Hill, J., Welsh, M., Brewer, E., Culler, D.: Tinyos: An operating system for sensor networks. In: Weber, W., Rabaey, J.M., Aarts, E. (eds.) Ambient Intelligence, pp. 115–148. Springer, Heidelberg (2005), doi:10.1007/3-540-27139-2_7
6. Of, M.N., Abrach, H., Carlson, J., Dai, H., Rose, J., Sheth, A., Shucker, B., Han, R.: Mantis: System support for. In: 2nd ACM International Workshop on Wireless Sensor Networks and Applications (WSNA), pp. 50–59 (2003)

GinConf: A Configuration and Execution Interface for Wireless Sensor Network in Industrial Context

José Cecílio, João Costa, Pedro Martins, and Pedro Furtado

University of Coimbra,
Coimbra, Portugal
jcecilio@dei.uc.pt, jpcosta@dei.uc.pt,
pmom@student.dei.uc.pt, pnf@dei.uc.pt

Abstract. Wireless sensor networks (WSNs) are deployed to sense, monitor and act on the environment. Some applications of these networks, especially in industrial sense and react scenarios, require high performance and guaranties. Our work is focused on building a system that allows configuring/reconfiguring alarms, actions or closed-loop techniques in the context of GINSENG project – wireless sensor networks with performance control guarantees. We propose an approach for interaction with real-world devices through a web services interface, allowing users to configure and apply various operations, including complex closed-loop techniques that monitor and act over any actuator in the WSN. To allow the interaction between a client application and the motes we implemented an API to access services of the motes.

Keywords: WSN, Configuration/Reconfiguration, Industrial application.

1 Introduction

Sensor Networks are used nowadays in many application contexts, with quite different characteristics. One application scenario of these networks consists on industrial environments. In an industrial setting for monitoring-and-control applications, easy configuration and reconfiguration capabilities become important during deployment and tests, where issues such as latencies may dictate modification. During the lifetime of the network, a deployment may not meet all requirements. It is important to check if all requirements are guaranteed, if not, it is needed to change something in the network.

Closed-loop control is an important issue in industrial settings. Since sensor motes have limited computation capability and control computations may require operation on values coming from multiple sensors, immediate sensor-triggered control will typically be only for emergency actuation (e.g. opening a valve if pressure goes beyond an emergency level). More complex closed-loop control computations can be done in a control workstation, subject to larger latencies and more data (e.g. multiple samples, inputs from multiple sensors).

In this paper, we present an approach to connect wireless sensor or actuator nodes to a web services interface for configuring and applying various operations, including complex closed-loop control techniques. This work is done in the context of a European GINSENG project – wireless sensor networks with performance control

P.J. Marron et al. (Eds.): REALWSN 2010, LNCS 6511, pp. 190–193, 2010.

guarantees. Sensors and actuators are represented as resources of the corresponding node and are made accessible using a web service interface that establishes the communication with the nodes. Our main goal is to enable a flexible architecture where sensor networks data can be accessed by users to configure the system, including configuration of alarms, sending rates, closed-loop control and actions. We present our system architecture and show a user interface called GWeb.

2 GinConf System Architecture

As devices of WSN have limited computation capabilities and may connect with different sensors and actuators, manual configuration/reconfiguration (by programming) is infeasible if a large number of sensor nodes are used. Based on web services, we introduce a plug-and-play approach, that allows configuring/ reconfiguring any mote in the WSN.

We designed the GinConf to offer the configuration and execution interface for the wireless sensor network. Our approach also allows connecting multiple concurrent applications to share sensing resources in a flexible way. Figure 1 shows the system architecture.

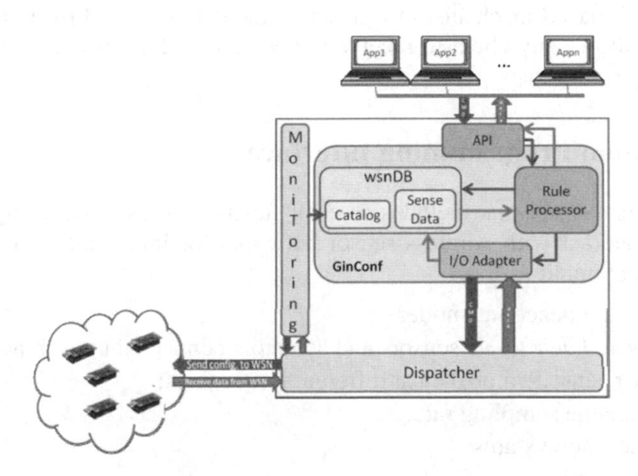

Fig. 1. System Architecture

As shown on Figure 1, we propose to use GinConf as an interface that allows configuring and executing controllers in wireless sensor networks.

GinConf abstracts the proprietary communication protocols of motes and offers their functionalities through an Application Programming Interface (API). It is based on a web service interface that allows connecting any applications to the WSN. For instance, if we consider an action coming from the closed-loop software via API to a sensor node, GinConf maps this request to a specific request for the mote and transmits it to the WSN. The API provides a set of resources that can be identified using a request of web service.

The architecture's key components of GinConf are the I/O Adapter, wsnDB and the Rule Processor. The I/O Adapter is a module that allows establishing the connection between GinConf and the dispatcher to obtain sensor data streams, submit data queries to the sensors, or access sensor characteristics. The dispatcher implements sensor-specific methods to communicate with the sensor.

The wsnDB is an internal memory database used to store information about the network. This module is subdivided in two sub-modules: catalog, and senseData. The catalog is responsible for indexing the sensor characteristics and other shared resources in the system, in order to enable applications to discover what is available for use. The catalog information is maintained up-to-date by the Monitoring module that collects status information from the network. To guarantee that the catalog has up-to-date information, the sensor node may periodically send status messages to the monitoring module. The senseData stores data messages during a time window. When an application needs data from overlapping space–time windows, senseData uses stored messages to get the data. This allows a client to request data from any past instant. For instance, if a client requests all values in the last 10 minutes to compute an appropriate action, GinConf extracts data from the memory, streams it and sends it to the application.

Lastly, the Rule Processor is a module that allows establishing the interaction between different clients over the network in order to exchange data or to trigger certain actions. The Rule Processor is composed by a web service interface that allows using the resources of remote devices. In addition, GinConf provides, through the API, a push-based mechanism to subscribe the data received from the WSN. This functionality allows any client to receive periodically a data stream with the readings transmitted by each sensor.

3 Application Programming Interface

GinConf offers an API that includes a set of functionalities required by applications that need to interact with wireless sensor networks for industrial environments. The API offers functionalities for:

- Activate / deactivate nodes;
- Activate / deactivate sensors and actuators connected to each node;
- Gather sensed value data at different frequencies;
- Change the sampling rate;
- Request node status;
- Reset a node;
- Change node configuration parameters;
- Send controller code to nodes;
- Start a controller;
- Stop a controller ;
- Set parameters, allow changing parameters of a controller;
- Define actions, conditions and rules.

4 User Interface

In this section, we describe the implementation of GWeb, an interactive application built on top of GinConf. GWeb (Figure 2) demonstrates how to create and send queries to the WSN.

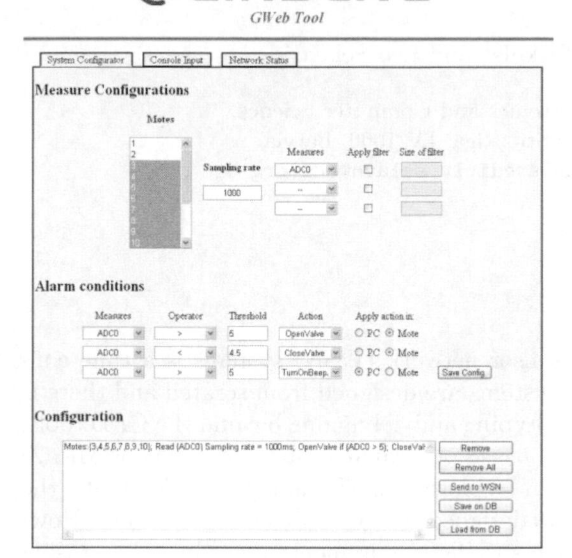

Fig. 2. GWeb Interface

Fig. 3. Layout of deployment

Fig. 4. Temperature's chart

This application allows users to, for example, define a set of rules which trigger certain actions based on a specified event. A typical rule may have the format *"if the pressure level in sensor A is greater than 5 bar, open valve X and send a notification to server"*.

GWeb is an application that combines configuration and display of sensor streams obtained using GinConf. Figure 3 shows the layout of an example of deployment and Figure 4 shows the chart of temperature for mote 3. This application can be used by all deployers to configure or reconfigure the network.

5 Demo Roadmap

In this demo we aim to present a tool that allows configuring and executing controllers in a WSN specifically designed for industrial scenarios. In the scope of the GINSENG Project and based on the architecture presented above, we aim to demonstrate how to configure the network to operate in critical scenarios. We will deploy some nodes in a tree hierarchy topology, where all nodes sense the temperature and we will demonstrate how to change the sampling rates, create alarms in motes and/or in PC, how to trigger actions based on events and how to change a threshold level.

EdiMote: A Flexible Sensor Node Prototyping and Profiling Tool

Rinalds Ruskuls[1] and Leo Selavo[2]

Institute of Electronics and Computer Science,
14 Dzerbenes Str, Riga, LV 1006, Latvia
rinalds.ruskuls@edi.lv, selavo@acm.org

1 Introduction

Designing hardware for wireless sensor network (WSN) systems is a time consuming process. Quite often new systems are designed from scratch and there is lack of support for the design, prototyping and debugging beyond the CAD tools. Therefore, many WSN systems are based on available platforms such as MicaZ, TelosB, EPIC [1] [7] [4]. Although these platforms offer a degree of flexibility, the hardware setup is limited to the predesigned sensor or extension modules. However, many WSN applications have specific requirements regarding the sensing types and fidelity while are very sensitive to the current draw by these sensors and waste by the unused components. Therefore, we propose a WSN hardware prototyping test bench EdiMote developed with the flexibility of prototyping, performance monitoring and hardware and software debug assistance in mind (Figure 1).

Our approach offers fast prototyping for virtually any hardware. The users new prototype is made from modules that are microcontrollers, communications transceivers, sensors, storage, any other electronic components or the combination of them. Each module is on a board with predefined interface with analog and digital signals and is either available from a hardware library of modules or custom designed by the user for the specific task. The boards are interconnected using a configurable digital and analog interconnect that allows fast reconfiguration, monitoring of the signals for debugging purposes and power consumption monitoring per module.

In addition, the EdiMote bench is capable of emulating any of the existing hardware such as Tmote Sky or MicaZ with the additional benefit of low level monitoring and signal debugging support. For example, an analog sensor value that is going to the ADC port of a controller can also be monitored and logged by the test bench along with the digital signals for profiling or a scenario replay purposes. Once the new system is tested, profiled and debugged enough, the user may proceed to create a standalone version of the unit from the final version of the hardware configuration where most problems have been resolved using the EdiMote prototyping tool.

This Demo intends to show the capabilities and usefulness of the EdiMote prototyping and profiling system. We show the tool in action, using various

P.J. Marron et al. (Eds.): REALWSN 2010, LNCS 6511, pp. 194–197, 2010.

Fig. 1. EdiMote prototyping and profiling board

modules and configurations, including the emulation of a TelosB mote or another well known sensor node.

2 System Architecture

The high level block diagram of the test bench architecture is shown in Figure 2.

Main idea of this system is to use reconfigurable analog and digital interconnect between the Test MCU and Modules. Note, that the system does not restrict to a single MCU, but can have multiple MCUs on several modules. The interconnect configuration is loaded from a computer via USB port. The Test MCU module is a daughter board with 80 pins for power, digital and analog signals. Several MCU daughter boards are available and more can be designed by the user, for example, based on MSP430, Atmel, Nordic Semiconductors and ARM controllers. The Modules typically have sensors, RF transceiver, extra controller or other application specific devices on them. The system supports up to

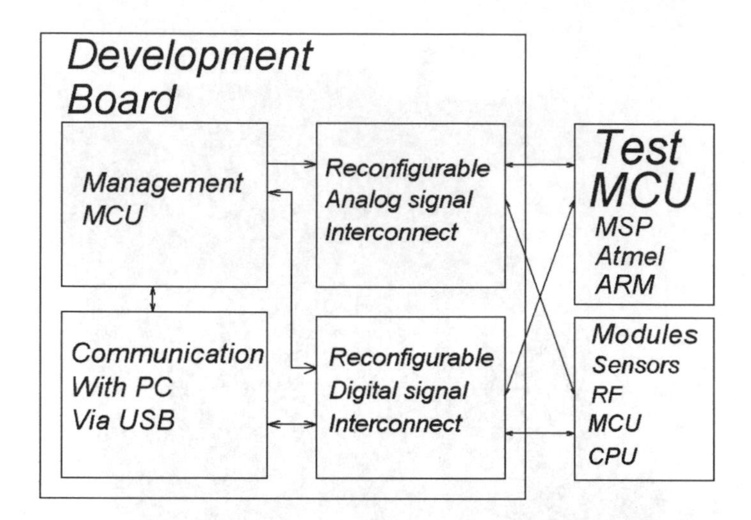

Fig. 2. Development board architecture

three modules with two supporting 4 analog and 10 digital signals, while the third module connector has 8 analog and 22 digital signals. Sensor modules like RF and MCU can be connected to those connectors.

2.1 Implementation

The digital signal matrix interconnecting GPIO and programming signals including JTAG is implemented with Altera MAXII EPM1270T144C5N CPLD [2] . The interconnect can perform digital level translation for systems where different levels are used. In addition the CPLD implements the monitoring of digital lines and dynamic reconfiguration of the interconnect. Advanced user might even take advantage of the CPLD reprogramming to implement custom profiling or debugging functions. The CPLD can be controlled by USB interface (FT4232) or by the EdiMote management MCU (MSP430F5437). The analog signal interconnect is implemented using AD75019 analog matrix(AMAT), which can connect 16x16 analog signals [3]. The analog matrix supports signal voltage levels from 4.5V to 4.5V. It is configured by the management MCU via SPI interface. The analog channels can be monitored by the management MCU ADC channels using operational amplifiers for offset implementation and gain correction. The management MCU controls the whole development board (Figure 3).

Besides the functionality described before, the MCU controls the power supply distribution. Another feature is 4 channel energy consumption monitoring system, based on the SPOT design [6]. The power consumption is monitored as needed for each attached module and the whole system. EdiMote system communicates the configuration, debugging and profiling information to a computer when attached to USB port. The system also has a standalone operation mode while storing the profiling data to a SD memory card for offline processing. The USB communication is implemented using FT4232 device featuring 4 UART

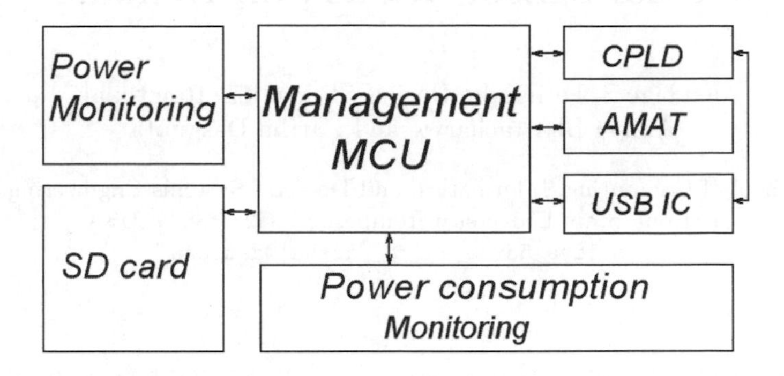

Fig. 3. EdiMote board management

ports operating as two UART/Bit-Bang ports plus two MPSSE engines used to emulate JTAG, SPI, I2C, Bit-bang or other synchronous serial modes [5]. Two JTAG interfaces are used for CPLD and in-system Test MCU programming. The management MCU is programmed via BSL. The last channel of the USB is used for CPLD configuration and data monitoring using Bit-bang interface. EdiMote software tool is used for the development board configuration and monitored data visualization.

References

1. MicaZ datasheet, `http://courses.ece.ubc.ca/494/files/MICAz_Datasheet.pdf`
2. Altera: MaxII Daasheet, `http://www.altera.com/literature/hb/max2/max2_mii5v1_01.pdf`
3. Analog Devices: AD75019 datasheet, `http://www.analog.com/static/imported-files/data_sheets/AD75019.pdf`
4. Dutta, P., Taneja, J., Jeong, J., Jiang, X., Culler, D.: A building block approach to sensornet systems. In: Proceedings of the 6th ACM Conference on Embedded Network Sensor Systems, SenSys 2008, pp. 267–280. ACM, New York (2008)
5. Future Technologies: FT4232H datasheet, `http://www.ftdichip.com/Documents/DataSheets/DS_FT4232H.pdf`
6. Jiang, X., Dutta, P., Culler, D., Stoica, I.: Micro Power Meter for Energy Monitoring of Wireless Sensor Networks at Scale. In: The Sixth International Conference on Information Processing in Sensor Networks (IPSN 2007) Track on Sensor Platforms, Tools, and Design Methods (SPOTS 2007), Berkeley, California, p. 10 (2007)
7. Moteiv: TelosB datasheet, `http://www.willow.co.uk/TelosB_Datasheet.pdf`

Virtual Sensor WPAN on Demand

Meddage S. Fernando, Harie S. Bangalore Ramthilak,
Amiya Bhattacharya, and Partha Dasgupta

School of Computing, Informatics, and Decision Systems Engineering,
Arizona State University, Tempe, AZ 85287-8809, USA
{saliya,harie,amiya,partha}@asu.edu

Abstract. Virtualization of wireless sensor PANs would be useful for general purpose networked sensing with application concurrency, for community-based sensor sharing, and for supporting platform heterogeneity as well as robustness. This abstract presents a snapshot of the preliminary design and implementation of a middleware for lightweight sensor network virtualization that makes use of the latest developments in TinyOS.

Keywords: Wireless PAN, multithreading, virtualization, TinyOS.

1 Introduction

Wireless sensor networks (WSN) are often deployed as mesh-connected wireless personal area networks (WPAN) of low-power sensor nodes (commonly known as "motes"), hanging off of an Internet gateway. WSN deployments are mostly application-specific, typically operated under a single administrative domain. Changing this norm, however, can bring forth interesting usage scenarios such as time-shared virtual WSN infrastructure [2] or sharing sensors across a community [3]. But these possibilities remained severely limited by initial OS design choices dictated by resource-poor hardware and the need for preserving battery power. Recent design breakthroughs in mote OS, such as preemptive multithreading and dynamic linking of program modules, bring forth lightweight WPAN virtualization towards provisioning these goals.

2 Virtual Sensor WPAN

Here we introduce a technique for spanning a virtual sensor WPAN formation on top of several existing host WPAN substrates. Earlier notions of virtual sensor network have restricted the virtual WPAN nodes to a proper subset of a larger substrate network, thereby creating smaller subnets for different dedicated applications [2]. Augmenting the spanning capability, our design even allows the virtual sensor network to grow even larger than any of its hosts so as to instantiate sensing applications work across domains.

P.J. Marron et al. (Eds.): REALWSN 2010, LNCS 6511, pp. 198–201, 2010.

2.1 Underlying Platform

Implementation of this middleware is done on TinyOS, an event-driven embedded operating system for networked sensing. TOS 2.x, the current TinyOS code base, provides the following vital components essential for forming this virtual WPAN over the hosting physical WPANs:

1. First, concurrency is one of the most important requirements for supporting more than one sensing/actuating application. Introduction of a fully preemptive threads package (TOSThreads) in TOS 2.1 has been leveraged to support concurrency in our system [4]. Each sensing application, including the native application of the host sensor networks, is treated as a thread. The TinyOS operating system itself runs as the high-priority kernel thread, which is liable for handling all tasks posted by application threads.
2. In order for running a different networked sensing application per virtual WPAN, the corresponding application thread must assume distinct network identities (a common PAN-id over the entire span of the virtual network, along with unique node-id within each WPAN). Capability of setting both PAN and node id's dynamically, a feature in rare use in the WSN research community, has however been available even in earlier versions of TinyOS.
3. To alleviate the burden of pre-loading sensing applications to motes and to preserve the flexibility of injecting new application modules on demand to virtual sensor networks, we also need a dynamic loading and linking mechanism for threads. We have used the new addition to the TinyOS library, namely TinyLD [5], which links the program code to the TinyOS kernel dynamically, either from the flash or the RAM.

2.2 Establishment of the Virtual WPAN

The formation of the virtual network is done on the respective host networks in the following three phases:

1. First a message for enabling a new PANID is broadcast to all nodes in the host network. While there are several protocol choices for this in TinyOS, our prototype uses DHV [1]. Once this PANID_ADDITION control packet reaches a node, it adds that PANID to its own PANID list. A node is supposed to listen to all the PANID's listed in its PANID list. Subsequently, a unicast message containing the newly assigned virtual node id is to be sent to each participating node.
2. In the second phase, application code modules are to be multicast to the motes selected to participate in the virtual WPAN. Since there is currently no efficient implementation of multicast over WPANs under TinyOS, our prototype assumes a dense participation, and thereby substitute multicast with a controlled flooding. Under sparse participation, it may be efficient to substitute it with multiple unicast routes. The middleware augments the TinyOS kernel thread to receive the packets destined to multiple network incarnations (combination of PAN and node id's) associated with the running

application threads, and demultiplexing to their respective input queues. On successful completion of the first two phases, each participating mote must send a cumulative acknowledgment (acknowledging the receipt of the code, virtual PANID and the virtual node id) to their respective gateways.

3. Finally, in the gateway selection phase, a virtual gateway has to be designated out of the participating host gateways. Our current implementation elects the host gateway that donates the largest number of motes to the virtual WPAN. Figure 1 provides a visual representation of how the networks look like after the formation of the virtual PAN. After the selection, the elected gateway sends a welcome message to all the nodes in the virtual WPAN (using the virtual WPAN's new PAN id). Gateway assignment can be changed on the fly, but we are yet to find the behavior of collection protocols (such as CTP) in response to that.

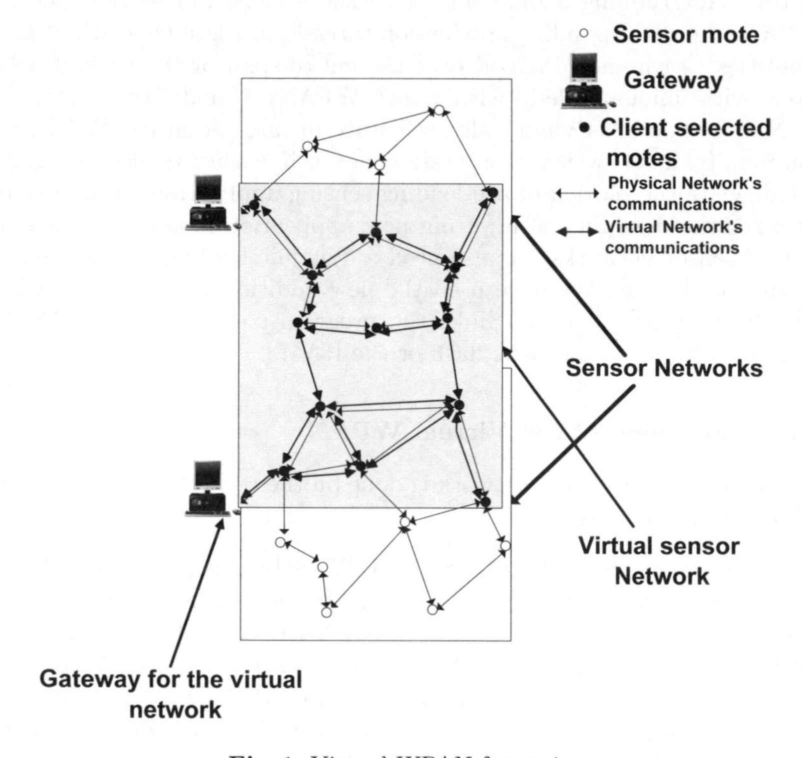

Fig. 1. Virtual WPAN formation

2.3 TinyOS Modifications

While the middleware is designed to run on top of TinyOS preserving all of its documented functionalities, a few changes to the existing TinyOS libraries were necessary. To be more precise, WPAN virtualization is implemented on a patched version of TinyOS. The TinyOS "receive" interface is one of the main parts in the library that needed to be patched. In the unmodified version, each wireless

PAN has a unique network id (called PANID) associated with it. This unique id allows each node on a WPAN to filter the packets so that a particular node can ignore the packets flowing in a different WPAN. Since the virtual WPAN in our design is a full-fledged WPAN in its own right, we must associate a unique PANID with it. By doing this, we are introducing new PANIDs in the host substrate network dynamically. We have changed the receive interface in such a way that a node listens to only a list of valid PANIDs instead of listening to just one.

3 Ongoing and Future Work

In its current stage of an initial prototype, the middleware for WPAN virtualization uses a simple controlled flooding protocol for incremental dissemination of application code modules. While the code dissemination is left outside the scope of the middleware by design, a robust and efficient code dissemination protocol is a complementary part of research in progress. In addition, being unaware of the fact that motes may spend their life in multiple incarnations, stock routing protocols are not expected to port onto virtual sensor WPANs without necessary modifications. Porting common routing protocols in TinyOS would follow once the middleware prototype becomes stable. Maintaining and secured sharing of code modules over the virtual WPAN nodes based on the tradeoff between trust and radio transmission cost constitutes another direction of future work.

Acknowledgment

This material is based upon work supported in part by the National Science Foundation under Grant No. CNS-1011931. Any opinions, findings, and conclusions or recommendations expressed in this material are those of the author(s) and do not necessarily reflect the views of the NSF.

References

1. Dang, T., Bulusu, N., Feng, W., Park, S.: DHV: A code consistency maintenance protocol for wireless sensor networks. In: Proceedings of the 6th European Conference on Wireless Sensor Networks (EWSN 2009), Cork, Ireland (February 2009)
2. Jayasumana, A.P., Han, Q., Illangasekare, T.: Virtual sensor networks: A resource efficient approach for concurrent applications. In: Proceedings of the 4th International Conference on Information Technology—New Generations (ITNG 2007), Las Vegas, Nevada, USA (April 2009)
3. Kansal, A., Nath, S., Liu, J., Zhao, F.: SenseWeb: An infrastructure for shared sensing. IEEE Multimedia 14(4), 8–13 (2007)
4. Klues, K., et al.: TOSThreads: Thread-safe and non-evasive preemption in TinyOS. In: Proceedings of the 7th ACM Conference on Embedded Networked Sensor Systems (SenSys 2009), Berkeley, California, USA (November 2009)
5. Musăloiu-E R., Liang, C.M., Terzis, A.: A modular approach for WSN applications. HiNRG Technical report 21-09-2008, Johns Hopkins University (2008)

TikiriAC: Node-Level Equally Distributed Access Control for Shared Sensor Networks

Nayanajith M. Laxaman, M.D.J.S. Goonatillake, and Kasun De Zoysa

University of Colombo School of Computing
No. 35, Reid avenue, Colombo 7, Sri Lanka
{nml,jsg,kasun}@ucsc.cmb.ac.lk
http://www.ucsc.cmb.ac.lk/wasn

Abstract. In this paper, we propose an access control mechanism that can be used to overcome challenges and problems related to access controlling in a shared Wireless Sensor Network (WSN) databases with complex connectivity topologies.

Keywords: Distributed Access Control, Shared Wireless Sensor Networks, Privilege Management Infrastructure, Public Key Infrastructure.

1 Introduction

Researchers and organizations from various disciplines are interested in using WSN for their research and applications. Deploying a sensor network of their own is a time consuming, infeasible, and a complicated task for companies and organizations such as universities, research groups, small business groups, and other interested individuals due to high cost of the devices, not authorized to deploy, etc. Therefore, the concept of Shared Wireless Sensor Networks (SWSN) is getting popular among these communities [1]. However, providing shared access for WSN has given rise to a different set of problems. An important issue which we consider in this paper is controlling access among SWSN users.

A considerable amount of research has been carried out in the area of controlling access of users within a SWSN. There have been mainly four approaches found in related research literature for authentication and authorization of users for SWSNs. 1) Centralized, 2) Selectively distributed within SWSN, 3) Equally distributed within SWSN, 4) Client side [2], [3], [4]. However, there are pros and cons of each of these approaches depending on the topology used to access the SWSN. For example, access controlling measures of a SWSN with single entry point would be different from the measures considered in a SWSN with multiple entry points. Therefore, it is challenging to come up with a solution that can address the issues which would arise in any SWSN topology. In this paper we propose a solution which addresses all these SWASN topologies. TikiriDB is a database abstraction which enables sharing sensor network whilst supporting all these topologies of user connectivity [1]. Therefore, we developed our solution as a module to the TikiriDB.

P.J. Marron et al. (Eds.): REALWSN 2010, LNCS 6511, pp. 202–205, 2010.

2 Our Approach

Since, a particular user may have the total control over the client application, client side authentication and authorization would be the least preference when giving a solution to the mentioned problem. Centralized approach has the limitation of single point of failure. Therefore, a distributed access controlling mechanism is preferable where the failure of several SWSN nodes may have a limited impact on total access controlling system. However, if access controlling has been distributed to handle individually by the nodes themselves, probability of failure can further be reduced. Therefore, in our proposed access controlling solution for SWSN, we opted to handle access controlling at node level, individually. In our approach, we opted to use public key certificates and attribute certificates to implement authentication and authorization in SWSN. Researchers have successfully implemented public key infrastructures on top of WSN using Elliptic Curve Cryptography (ECC) [5]. ECC scheme provides 1024 bit RSA equivalent security only by using 160 bit certificates. Therefore, many researchers in sensor network discipline have opted ECC as their primary cryptographic system [5], [6], [2]. Public key cryptography is used for initial secure communication and a shared key is exchanged between source and destination nodes to continue further communication using symmetric key cryptography.

2.1 TikririAC Module for TikiriDB

TikiriAC is developed as a module for TikiriDB. TikiriAC module has two main components where one is at the node and other is with the user. Enabling TikiriAC in TikiriDB pass the optimized query generated by TikiriDB client to TikiriAC. Then, TikiriAC handles authentication and authorization of users. The result of successful authorization process passes requested query to TikiriDB query processor at nodes. TikiriAC also ensures the security of the query results when they transferred back to the user.

2.2 TikiriAC Public Key Infrastructure

Figure 1 illustrates proposed public key infrastructure in TikiriAC. Public key certificates are issued by a Certification Authority (CA) and attribute certificates are issued by an Attribute Authority (AA). However, considering the resource limitation in sensor nodes we propose using common certificates for both AA and CA by forming a Hybrid Authority (HA). Furthermore, since network communication consumes a considerable amount of power of a sensor node, we further reduced the size of attribute certificate and public key certificate by removing several attributes from the certificates to reduce number of data packets propagated in WSN when initializing the security algorithm. Therefore, it should be mentioned that the certificates used in TikiriAC are not fully compliant with X.509 standard. In addition to that, for the time being, certificate revocation protocols have not been incorporated. Hence, the user certificate expiration time has been set to a very short period to make sure users renew their

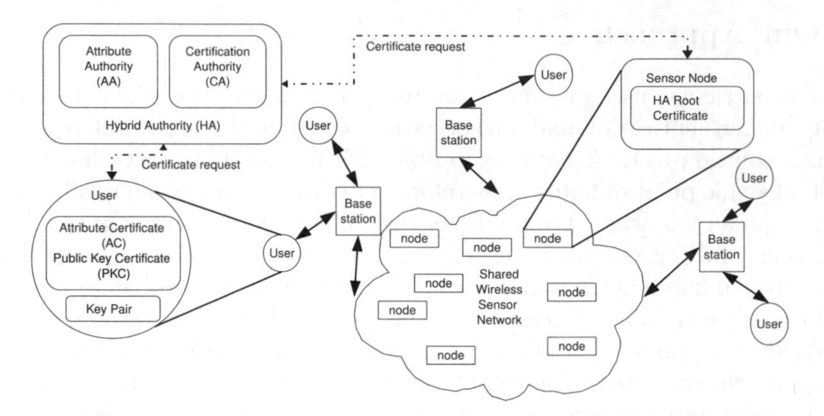

Fig. 1. TikiriAC public key infrastructure

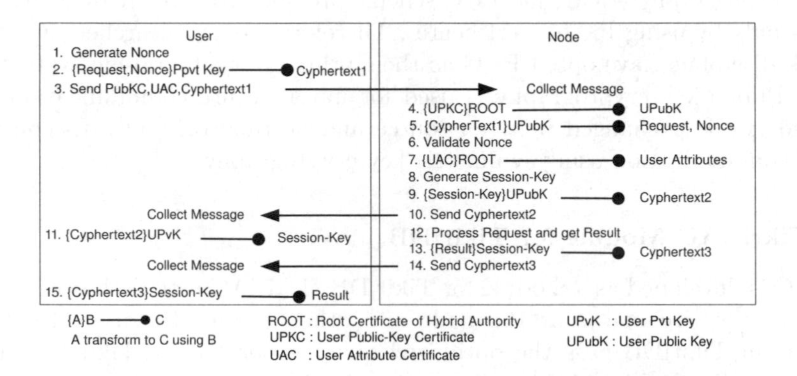

Fig. 2. TikiriAC Protocol

certificates frequently. Public key certificates are used to authenticate the users and to share a symmetric key between data requesting user and destined sensor nodes. Attribute certificates are to manage authorization of the users [7]. Every authenticated user of the SWSN has a public key certificate and an attribute certificate issued by the owner of the SWSN or a trusted coordination party. The public key certificate of the hybrid authority (HA root certificate) is burned in to each node at the time of deployment.

2.3 TikiriAC Protocol

Figure 2 illustrates TikiriAC protocol. First, the user who accesses the SWSN generates a nonce (a time stamp) for the current message to be sent with the message (1). The query and nonce is signed using requesters private key (2). The content is then sends to the destination node with the signature, and requesters public key certificate and attribute certificate (3). At the destination node, to verify the authenticity of the requester, node use ECC algorithms to verify requesters public key certificate using HA's root certificate (4). Then the public key certificate of the user is used to decrypt the encrypted content and reveals nonce and the query (5). Newly received nonce is used to prevent replay attacks (6). If the message possess a valid nonce, the attribute certificate is verified using the HA's root certificate (7). A valid attribute certificate is providing the accessibility constrain information for a particular user such as; which sensors the user can access, how long he can execute a query, what is the maximum frequency that the user can obtain information, etc. Within the TikiriAC protocol, if the message fails at any step of verification or validation, the request is discarded. After successfully completing the above process, node generates a session key to be used for the communication between the user and node itself (8). The session key is then encrypted with users public key and sends back to the user (9). Then the user decrypts the encrypted session key by using his/her private key and keeps the session key until this query execution finishes (10). Any further communication or new session key exchange is done through an encrypted channel between the user and the node. It should also be mentioned that it is required to encrypt the messages in certain situations such as for in-network aggregation of sensor data. For example, calculating the average temperature of given set of nodes. A group key generation and manipulation algorithm is introduced to overcome this issue.

3 Conclusions

Here, we have introduced a solid architecture to overcome the access control problem arising in shared sensor networks with complex topologies. High security was guaranteed in the use public key cryptography. We considered several measures to reduce the resource consumption caused due to public key cryptography in the sensor network. Finally we explained the appropriate architecture and technologies to implement our design as a module for TikiriDB.

References

1. Laxaman, N.M., Goonatillake, M.D.J.S., Zoysa, K.D.: Tikiridb: Shared wireless sensor network database for multi-user data access (2010)
2. Wang, H., Sheng, B., Li, Q.: Elliptic curve cryptography-based access control in sensor networks. Int. J. Security and Networks
3. Benenson, Z.: Authenticated queries in sensor networks. In: Molva, R., Tsudik, G., Westhoff, D. (eds.) ESAS 2005. LNCS, vol. 3813, pp. 54–67. Springer, Heidelberg (2005)
4. Networks, W.S., Karlof, C.: Tinysec: A link layer security architecture for wireless sensor networks
5. Liu, A., Ning, P.: Tinyecc: A configurable library for elliptic curve cryptography in wireless sensor networks
6. Gupta, V., Wurm, M., Zhu, Y., Millard, M., Fung, S., Gura, N., Eberle, H., Shantz, S.C.: Sizzle: A standards-based end-to-end security architecture for the embedded internet. Technical report (2005)
7. Johnston, W.: Authorization and attribute certificates for widely distributed access control (1998)

Author Index

Printing: Mercedes-Druck, Berlin
Binding: Stein+Lehmann, Berlin